Maintenance Management and Regulatory Compliance Strategies

Terry Wireman, C.P.M.M.
6477 West Buckskin Road
Pocatello, ID 83201
www.TerryWireman.com
TLWireman@mindspring.com

Industrial Press

Library of Congress Cataloging-in-Publication Data

Wireman, Terry.
Maintenance management and regulatory compliance strategies / Terry Wireman.
p. cm
Includes index.
ISBN (invalid) 0-8311-3217-6
1. Industrial safety--Law and legislation--United States. 2. Industrial safety--United States. 3. Plant maintenance--United States. I. Title

KF3570.W57 2003
344.73'0465--dc21

2003048682

Maintenance Management and Regulatory Compliance Strategies

Industrial Press Inc.
200 Madison Avenue
New York, New York 10016

Dedication

This book is dedicated to my wife Kay,
whose tireless patience
allowed me to complete this text.
May she live long and prosper

Table of Contents

Preface

Maintenance Management and Regulatory Compliance Strategies is a textbook designed to create an awareness of how a maintenance management function for a company can contribute or detract from a company's compliance with regulatory requirements. The book is not designed to provide a complete listing of every possible regulatory requirement and How maintenance managers can impact these requirements. The book is designed to give an over-view of each of the four major compliance programs. It is also going to provide a cross-section of regulations for each program that maintenance may impact including some little-known regulations of which most companies will find themselves in violation.

The four major areas the book addresses are:

OSHA
EPA
FDA
ISO-9000

OSHA

OSHA deals with the health and safety of workers in the workplace. Many OSHA regulations are written directly for individuals in maintenance-related departments within a company. A majority of the book will be spent considering these regulations. Again, the book does not guarantee that managers who follow the directions would be considered to be in compliance with all OSHA regulations.

What is OSHA?

OSHA *(WWW.OSHA.Gov)*, is a government office of the United States with the following mission statement:

The mission of the Occupational Safety and Health Administration (OSHA) is to save lives, prevent injuries and protect the health of America's workers. To accomplish this, federal and state governments must work in partnership with the more than 100 million working men and women and their six and a half million employers who are covered by the Occupational Safety and Health Act of 1970.

OSHA Services

OSHA and its state partners have approximately 2100 inspectors, plus complaint discrimination investigators, engineers, physicians, educators, standards writers, and other technical and support personnel spread over more than 200 offices throughout the country. This staff establishes protective standards, enforces these standards, and reaches out to employers and employees through technical assistance and consultation programs.

OSHA's Domain

Nearly every working man and woman in the nation comes under OSHA's jurisdiction (with some exceptions such as miners, transportation workers, many public employees, and the self-employed). Other users and recipients of OSHA services include occupational safety and health professionals, the academic community, lawyers, journalists, and personnel of other government entities. *

Note: While the OSHA standards (29 CFR) encompass regulations from part 70 – 71, part 1900 through 2400, the OSHA section of this book will focus on part 1910, where the majority of the general requirements are contained.

EPA

The mission of the U.S. Environmental Protection Agency is to protect human health and to safeguard the natural environment—air, water, and land—upon which life depends. *(www.epa.gov)*
EPA's purpose is to ensure that:

— All Americans are protected from significant risks to human health and the environment where they live, learn, and work.

—National efforts to reduce environmental risk are based on the best available scientific information.

—Federal laws protecting human health and the environment are enforced fairly and effectively.

—Environmental protection is an integral consideration in U.S. policies concerning natural resources, human health, economic growth, energy, transportation, agriculture, industry, and

*The above material is derived from documentation posted on the OSHA website.

international trade, and these factors are similarly considered in establishing environmental policy.

—All parts of society—communities, individuals, business, state and local governments, tribal governments—have access to accurate information sufficient to participate effectively in managing human health and environmental risks.

—Environmental protection contributes to making our communities and ecosystems diverse, sustainable, and economically productive.

—The United States plays a leadership role in working with other nations to protect the global environment.

EPA deals with pollutants that may be released during a plant's operation. There are many regulatory requirements for maintenance related to record-keeping and service of environmentally sensitive equipment. This book will highlight these regulations. Again, the book is not so all-inclusive as to contain every possible regulation. It will, however, presents a good cross-section of regulations that highlight the requirements for maintenance in the majority of areas.

FDA

The Food and Drug Administration (www.fda.org) touches the lives of virtually every American every day for it is FDA's job to examine the following:

food
cosmetics
medicines
medical devices
radiation-emitting products
feed and drugs for pets and farm animals

First and foremost, FDA is a public health agency, charged with protecting American consumers by enforcing the Federal Food, Drug, and Cosmetic Act and several related public health laws.

A primary mission of the Food and Drug Administration is to conduct comprehensive regulatory coverage of all aspects of production and distribution of drugs and drug products to assure that such products meet the 501(a)(2)(B) requirements of the Act. FDA has developed two basic strategies:

1) evaluating through factory inspections, including the collection and analysis of associated samples, the conditions and practices under which drugs and drug products are manufactured, packed, tested, and held, and

2) monitoring the quality of drugs and drug products through surveillance activities such as sampling and analyzing products in distribution.

This compliance program is designed to provide guidance for implementing the first strategy. Products from production and distribution facilities covered under this program are consistently of acceptable quality if the firm is operating in a state of control. The Drug Product Surveillance Program (CP 7356.008) provides guidance for the latter strategy.

The inspectional guidance in this program is structured to provide for efficient use of resources devoted to routine surveillance coverage, recognizing that in-depth coverage of all systems and all processes is not feasible for all firms on a biennial basis. It also provides for follow-up compliance coverage as needed.

FDA also ensures that all of these products are labeled truthfully with the information that people need to use them properly.

FDA investigators and inspectors visit more than 15,000 facilities a year, seeing that products are made safely and labeled truthfully. As part of their inspections, they collect about 80,000 domestic and imported product samples for examination by FDA scientists or for label checks.

If a company is found violating any of the laws that FDA enforces, FDA can encourage the firm to voluntarily correct the problem or to recall a faulty product from the market. A recall is generally the fastest and most effective way to protect the public from an unsafe product.

When a company can't or won't correct a public health problem with one of its products voluntarily, FDA has legal sanctions it can bring to bear. The agency can go to court to force a company to stop selling a product and to have items already produced seized and destroyed. When warranted, criminal penalties—including prison sentences—are sought against manufacturers and distributors.

Assessing risks—and, for drugs and medical devices, weighing risks against benefits—is at the core of FDA's public health protection duties. By ensuring that products and producers meet certain standards, FDA protects consumers and enables them to know what they're buying. For example, the agency requires that drugs—both prescription and over-the-counter—be proven safe and effective.

In deciding whether to approve new drugs, FDA does not itself do research, but rather examines the results of studies done by the manufacturer. The agency must determine that the new drug produces the benefits it's supposed to provide without causing side effects that would outweigh those benefits.

FDA deals with equipment that is used to preclude produce food beverage and pharmaceutical products. There are many regulations that focus on the equipment used in manufacturing processes. The book will highlight many of these regulations. Again, this book is not all-inclusive. Instead, a cross-section of regulations will be presented so that individuals could examine their company's compliance.

ISO-9000

In 1987, The International Standards Organization first published the standards on quality assurance, known as the ISO-9000 standards. There are three main models for the standards:

ISO-9001 – This standard is for companies whose conformity to specified requirements is met throughout the whole cycle from product design to product service. This is the most complete and stringent of the standards.

ISO-9002 – This standard is for companies that want to demonstrate their conformity to standards in production and installation.

ISO-9003 – This standard is for companies that need to demonstrate their capabilities in the inspection and testing, where the product is specifically for those requirements.

ISO-9000 regulations deal with quality and how equipment can impact quality. Many calibration and inspection items are required for maintenance. This book will highlight areas in which maintenance compliance is required to produce a quality product and meet the quality standards. Once again, the book is not so all-inclusive as to contain all the regulations. However, a good cross-section of regulations will be examined so that individuals can help assure their company's compliance.

Summary

The regulatory agencies described here are the most commonly encountered in companies in the United States. While some international companies are not bound by the regulatory agencies indigenous to the United states, they will typically have corresponding regulations in their particular country.

It is hoped that by understanding the standards in the context of maintenance (asset) management, compliance with these regulations will be easier to achieve and maintain.

Introduction to Maintenance Management

Functional Maintenance Best Practices and Regulatory Compliance

In today's business environment, terms like best practices, benchmarking, and world class continuously bombard us. However, what are best practices? What are common functional best practices for maintenance or asset management that you can use to benchmark other companies in order to achieve best practices within your own organization? How do you know if you have achieved world class status?

Best Practices Provide Competitive Advantage

A definition of best practices might begin with "the practices that enable a company to achieve a competitive advantage over its competitors in a specific business process." If that definition is adapted to the maintenance process, it would read: "The maintenance practices that enable a company to achieve a competitive advantage over its competitors in the maintenance process."

What are the functional best practices in maintenance management?

1. Preventive Maintenance
2. Inventory and Procurement
3. Work Flow and Controls
4. Computerized Maintenance Management System Usage
5. Technical and Interpersonal Skills
6. Operational Involvement
7. Predictive Maintenance
8. Reliability Centered Maintenance
9. Total Productive Maintenance
10. Financial Optimization
11. Continuous Improvement

THE MAINTENANCE MANAGEMENT PYRAMID

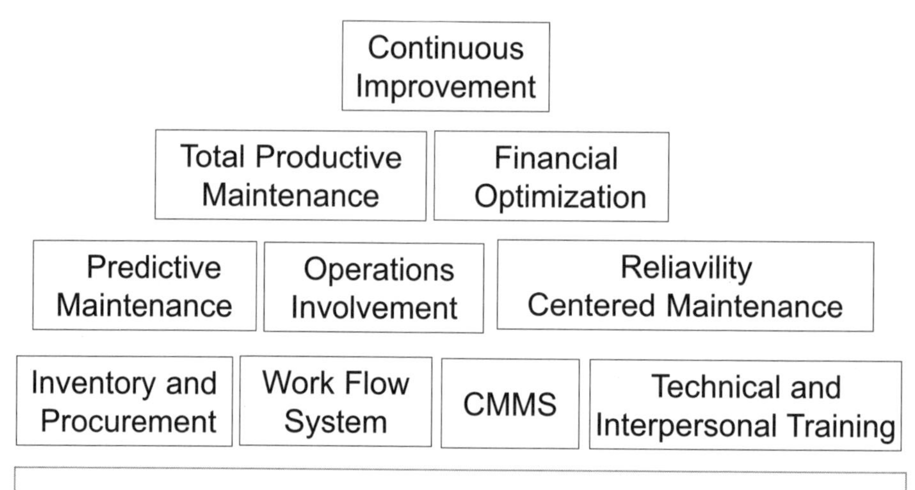

Figure 1 – The Maintenance Management Pyramid

The pyramid illustrates the structural relationship between functional best practice disciplines for maintenance/asset management.

1. Preventive Maintenance

The preventive maintenance (PM) program is the key to any attempt to improve the maintenance process. This program reduces the amount of reactive maintenance to a level that allows the other practices in the maintenance process to be effective. However, most companies in the United States have problems keeping the PM program focused. In fact, surveys have shown that only 20 percent of U.S. companies think their PM programs are effective.

Most companies need to focus on the basics of maintenance if they are to achieve any type of best-in-class status. Effective PM activities would enable a company to achieve a ratio of 80 percent proactive maintenance to 20 percent (or less) reactive maintenance. Once the ratios are at this level, other practices in the maintenance process become more effective.

When it comes to the issue of regulatory compliance, the preventive maintenance program is the single most essential function. The vast majority of regulations, whether for OSHA, EPA, FDA, or ISO, require maintaining assets in prime condition. In addition, the documentation necessary to achieve and maintain compliance is mainly a function of the preventive maintenance program.

2. Inventory and Procurement

The inventory and procurement programs must focus on providing the right parts at the right time. The goal is to have enough spare parts without having too many spare parts. However, the interdependency between the practices becomes apparent: No inventory and procurement process can cost-effectively service a reactive maintenance process. However, with the majority of maintenance work planned several weeks in advance, the practices within the inventory and procurement process can be optimized.

What level of performance is typical in companies today? Many companies see service levels below 90 percent, which means stockouts run greater than 10 percent of requests made. This level of service leaves customers (maintenance personnel) fending for themselves, stockpiling personal stores and circumventing the standard procurement channels to obtain their materials. They do not do this for personal reasons, but rather they want to provide service to their customers (operations or facilities). It is really a self-defense mechanism.

To prevent this situation, companies must institute the type of stores controls that will allow the service levels to reach 95 to 97 percent with 100 percent data accuracy. When this level of stores and procurement performance is achieved, you can then start the next step toward improvement.

These controls are imperative if all spare parts are to be tracked and controlled. Most regulatory programs require the tracking of spare parts and other materials that may have an impact on the integrity of the equipment or asset. In some cases, such as rebuildable spares, the tracking of the component is crucial to insure equipment/asset integrity. Such tracking requires tracing

Note – Figures One and Two of this introduction are taken from Developing Performance Indicators for Managing Maintenance Wireman, Terry. Industrial Press: New York City, New York. September 1998. ISBN 0831130806.

information not only about a location, but also about each individual component. Seemingly small details like this require extensive documentation and recordkeeping on the part of a maintenance organization.

3. Work Flows and Controls

This practice involves documenting and tracking the maintenance work that is performed. A work order system is used to initiate, track, and record all maintenance activities. The work may start as a request that needs approval. Once approved, the work is planned, then scheduled, performed, and finally recorded. Unless the discipline is in place and enforced to follow this process, data is lost, and true analysis can never be performed.

The solution requires comprehensive use of the work order system to record all maintenance activities. Unless the work is tracked from request through completion, the data is fragmented and useless. If all of the maintenance activities are tracked through the work order system, then effective planning and scheduling can start.

Planning and scheduling requires someone to perform the following activities:

Review the work submitted

Approve the work

Plan the work activities

Schedule the work activities

Record the completed work activities

Unless a disciplined process is followed for these steps, productivity decreases and reduced equipment downtime never occurs. Such results leave the perception that maintenance planning is a clerical function, making it vulnerable to the first cuts when any type of reduction in overhead costs are examined. At least 80 percent of all maintenance work should be planned on a weekly basis. In addition, the schedule compliance should be at least 90 percent on a weekly basis.

As respects the regulatory requirements, the work order function is crucial to recording all work that may have an impact on the equipment/assets. This documentation will include the date and time the work was performed and the detail of how the work was performed down to identifying the particular maintenance technician who performed the work and the skill level of the technician. These records may be kept manually, but in most cases are stored electronically in the computerized maintenance management system.

4. Computerized Maintenance Management Systems Usage

In most companies, the maintenance function utilizes sufficient data to require the computerization of the data flow. This facilitates the collection, processing, and analysis of the data. The usage of the Computerized Maintenance Management System (CMMS) has become popular in most countries around the world. CMMS software manages the functions discussed previously, and provides support for some of the best practices that will be mentioned in subsequent material.

CMMS has been used for almost a decade in some countries with very mixed results. A recent survey in the United States showed the majority of companies using less than 50 percent of their CMMS capabilities. This means the data collected by these companies is highly suspect and probably highly inaccurate. One requirement for a company to be effective in CMMS usage is complete usage of its system and complete accuracy of the data collected.

The lack of complete usage of a CMMS is a critical factor in recordkeeping compliance. The lack of complete and/or accurate recordkeeping requires that companies maintain a separate regulatory recordkeeping system. In lieu of the separate system, many companies will fail a check of their documentation by a regulatory agency. It is only by dedicating sufficient resources to the CMMS utilization that companies will have sufficient electronic documentation for compliance recordkeeping.

It should be noted that electronic data collection and reporting systems are acceptable to the regulatory agencies. This eliminates the need to have redundant paper systems in addition to electronic systems. In one of the responses to questions about electronic recordkeeping, OSHA responded:

> 29 CFR 1910.179(j)(2)(iii) and 1910.179(j)(2)(iv) requires "monthly inspection with a certification record which includes the date of inspection, the signature of the person who performed the inspection and . . ." If the employer's use of the

electronic system that identifies the person who performed the inspection, in lieu of the signature, is an acknowledgment that the person delegated responsibility for the inspection is certifying that it is true and complete to the best of that person's knowledge, then it would meet the intent of the record keeping requirements in 1910.179. In addition, an electronic system monthly inspection record which does not include the person's signature as required in 1910.179, would be considered a de minimis violation. De minimis violations are violations of standards which have no direct or immediate relationship to safety and health and will not be included in citations."

As will be shown in the chapter on Computerized Maintenance Management Systems, electronic work order systems are accepted by the regulatory agencies.

5. Technical and Interpersonal Training

This function of maintenance insures that the technicians working on the equipment have the technical skills that are required to understand and maintain the equipment. Additionally, those involved in the maintenance functions must have the interpersonal skills to be able to communicate with other departments in the company. They must also be able to work in a team or natural work group environment. Without these skills, there is little possibility of maintaining the current status of the equipment. Without these skills, the probability of ever making any improvement in the equipment is inconceivable.

While there are exceptions, the majority of companies today lack the technical skills within their organizations to maintain their equipment. In fact, studies have shown that almost 1/3rd of the adult population in the United States is functionally illiterate or just marginally better. When these figures are coupled with the lack of apprenticeship programs available to technicians, the specter of a workforce where the technology of the equipment has exceeded the skills of the technicians that operate or maintain it has become a reality. These issues pose difficulties for companies trying to comply with regulatory agencies.

6. Operational Involvement

The operations or production departments must take ownership of their equipment to the extent that they are willing to support the maintenance department's efforts. Operational involvement, which varies from company to company, includes some of the following activities:

— Inspecting equipment prior to start up

— Filling out work requests for maintenance

— Completing work orders for maintenance

— Recording breakdown or malfunction data for equipment

— Performing some basic equipment service, such as lubrication

— Performing routine adjustments on equipment

— Executing maintenance activities (supported by central maintenance)

The extent to which operations is involved in maintenance activities may depend on the complexity of the equipment, the skills of the operators, or even union agreements. The goal should always be to free up some maintenance resources to concentrate on more advanced maintenance techniques.

The impact that operations' involvement has on the regulatory compliance programs varies depending on the activities carried out by the operations personnel. Where the operations personnel are heavily involved, additional training will be required for them to become proficient at regulatory recordkeeping. Additional technical training will also be required for most of the personnel to carry out the work in a safe and compliant manner. This training will prevent violation of lock out/tag out regulations.

Extensive coverage of various training requirements for regulatory compliance is given in the chapter on Training and Skills Development.

7. Predictive Maintenance

Once the maintenance resources have been freed up because the operations department has become involved, these resources should be refocused on the predictive technologies that apply to their assets. For example, rotating equipment is a natural fit for vibration analysis, electrical equipment for thermography, and so on.

The focus is to investigate and purchase technology that solves or mitigates chronic equipment problems that exist, not to purchase all of the technology available. The predictive maintenance (PDM) inspections should be planned and scheduled utilizing the same techniques that are used to schedule the preventive tasks. All data should be recorded in or interfaced to the CMMS.

The predictive tools can do much to insure regulatory compliance. Instead of watching for visible signs of impending failure or replacing certain components on a scheduled basis, companies can now replace equipment components based on their actual condition. This eliminates the unnecessary expense of premature component replacement and yet still insures regulatory compliance as respects the component condition. Predictive maintenance can be a valuable tool for process safety management.

8. Reliability Centered Maintenance

Reliability Centered Maintenance (RCM) techniques are now applied to the preventive and predictive efforts to optimize the programs. If a particular asset is environmentally sensitive, safety related, or extremely critical to the operation, then the appropriate PM/PDM techniques are selected and implemented.

If an asset is going to restrict or impact the production or operational capacity of the company, then one level of PM/PDM activities is applied with a cost ceiling in mind. If the asset will be allowed to fail and the cost will be the expense of replacing or rebuilding the asset, then another level of PM/PDM activities is specified. There is always the possibility that it is more economical to allow some assets to run to failure and this option is considered in RCM.

The RCM tools require data to be effective. For this reason, the RCM process is used after the organization has progressed to the point that ensures accurate and complete asset data.

RCM is an essential part of any maintenance program if the cost of regulatory compliance is to be minimized without compromising the integrity of the program. Using RCM techniques for equipment in the design phase through to the actual operational phase will help to insure the equipment is designed for regulatory compliance and will maintain regulatory compliance through to the decommissioning phase of its life cycle. Most of the regulations related to design criteria will be covered in this section.

9. Total Productive Maintenance

Total Productive Maintenance (TPM) is an operational philosophy whereby all workers in the company understand that their job performance impacts the capacity of the equipment in some way. For example, operations may not understand the true capacity of the equipment and run it beyond design specifications, which could create unnecessary breakdowns. Another example is the purchasing department that always buys the spare parts to the correct specifications—not trying to save a small amount of money and creating breakdowns because the parts did not last as long as they should.

TPM is like Total Quality Management. The only difference is that companies focus on their assets, not their products. TPM can handle all of the tools and techniques used to implement, sustain, and improve the total quality effort.

TPM is enhanced when implemented as a part of regulatory compliance. With all employees focused on the condition of the company assets, regulatory compliance becomes routine. The level of effort to maintain regulatory compliance is minimized, due to the fact that many of the TPM activities are synergistic. This focus allows complete compliance at minimum cost.

There is no separate chapter on Total Productive Maintenance, since the regulatory requirements related to TPM are covered in other chapters.

10. Financial Optimization

This statistical technique combines all of the relevant data about an asset, such as downtime cost, maintenance cost, lost efficiency cost, and quality costs. It then balances that data against financially optimized decisions, such as when to take the equipment offline for maintenance, whether to repair or replace an asset, how many critical spare parts to carry, and what the maximum-minimum levels on routine spare parts should be.

Financial optimization requires accurate data, since making these types of decisions incorrectly could have a devastating effect on a company's competitive position. When a company reaches a level of sophistication where this technique can be used, it is approaching best-in-class status.

From a regulatory compliance standpoint, cost versus compliance is not a consideration. However, if the maintenance/asset management techniques employed to this point are properly focused, overall cost to the company will be minimized.

11. Continuous Improvement

Continuous improvement is best epitomized by the expression, "best is the enemy of better." Continuous improvement in asset care is an ongoing evaluation program. This includes constantly looking for the little things that can make a company more competitive.

Benchmarking is one of the key tools for continuous improvement. Of the several types of benchmarking practices, one of the most successful is process benchmarking, which examines specific processes in maintenance, compares the processes to companies that have mastered those processes, and maps changes to improve the specific process.

The key to benchmarking is self-evaluation. A company must know its current status before it tries to benchmark against other companies. Without this knowledge, it is impossible to obtain an accurate comparison of the benchmarked process.

Benchmarking is a very useful tool when it comes to regulatory compliance. Understanding how other companies achieve compliance can help companies modify their approach to achieve a higher level of compliance or reduce the current cost of compliance. By studying other companies' approaches, implementing improvements, and monitoring the improvements, increased compliance levels can be achieved.

Performance Measurement, Benchmarking, Best Practices and Regulatory Compliance

Performance indicators or measures for best practices are misunderstood and misused in most companies today. Performance indicators are just that, an indicator of performance. They should not be used for ego gratification; that is, to be used for benchmarking against another company to show how much better one company is than another. Performance measures are also not to be used to show "we are just as good as everyone else in our market, so we do not need to change."

Properly used, performance indicators should highlight opportunities for improvement within companies today. Performance measures should be used to highlight a soft spot in a company, further analyze to find the problem that is causing the indicator to be low, and then ultimately point to a solution to the problem. This process implies that there should be multi-level indicators.

One layer of indicators might be at a corporate strategic level. A supporting level would be the financial performance indicator for a particular department or process. A third level would be an efficiency and effectiveness indicator that highlights what impacts the financial indicator. A fourth level would be a tactical level indicator that highlights the departmental functions that contribute to the efficiency and effectiveness of the department. The fifth level of indicator is the measurement of the actual function itself, as highlighted by the ten best practices.

In an effort to help clarify this tiered approach to performance indicators, consider the pyramid in Figure 2. While the pyramid gives the hierarchical relationship of the performance indicators, it should be noted that the indicators are determined, not from the bottom up, but from the top down. The corporate indicators are measuring what is important to top management in order for the needs of the stakeholders or shareholders to be satisfied. So the corporate level indicators will help the organization focus their efforts on supporting the company direction.

Rule Number 1–All performance indicators must be tied to the long range corporate business objectives.

If a corporate indicator highlights a weakness, then the next lower level of indicators should give further definition and clarification to the cause of the weakness. When the functional performance indicator level is reached, the particular problem function should be highlighted. It will then be up to the responsible manager to take action to correct the problem condition. When the problem is corrected, the indicators should be monitored for improvement at the next vertical level to ensure the action taken was appropriate. If the appropriate action was taken, the improvement should be noticed as it impacts the hierarchical indicators up to the original corporate indicator.

The advantages of timeliness, utilizing hierarchical performance indicators, is clear when examining the improvement process. If changes made at the functional level do not result in a change in the

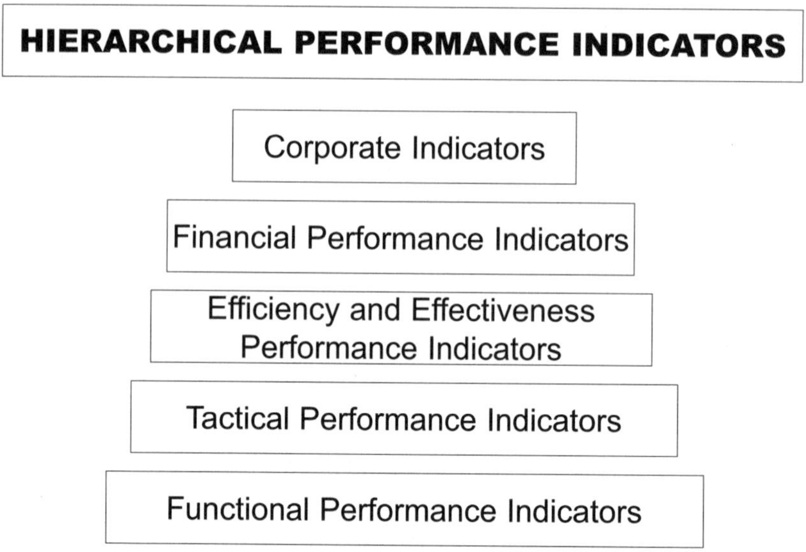

HIERARCHICAL PERFORMANCE INDICATORS

Corporate Indicators

Financial Performance Indicators

Efficiency and Effectiveness Performance Indicators

Tactical Performance Indicators

Functional Performance Indicators

Figure 2: This pyramid illustrates the relationship between the levels of indicators.

tactical performance indicator, then it is obvious the changes made were incorrect. This should be apparent quickly so the organization will not have to wait until the end of the month or the end of the quarter to evaluate the effect on the corporate indicator.

Selecting Performance Indicators to Support Company's Goals

The process of implementing new performance indicators so that they become ingrained in the culture of the business presents both opportunity and challenge. The opportunity is for each department to connect its operation to the overall business of the company. The challenge is to find indicators that allow these connections to be easily made.

The correct way to develop performance indicators is to work from the top, or corporate level, then develop indicators at each subsequent level and allowing the indicators to be connected. If the indicators are selected at the bottom and then built upward, they may be conflicting rather than supportive.

Of course, all levels of the pyramid will require compliance measurement. At a functional level, it might be something as simple as regulatory PM compliance measurement. At a tactical level, the measure could be resource allocation for compliance PMs, insuring that sufficient resources are allocated to complete all regulatory PMs. At an efficiency and effectiveness level, it might be how accurate the job plans are for the compliance PMs or how effective the PMs are at eliminating regulatory compliance infractions. At a financial level, the measure might be the financial impact of the regulatory compliance program. At the corporate level, it might be the level of compliance achieved, or even an intangible measure such as community perception of the plant, its safety, and its health records.

The bottom line: All compliance measures must be a part of the overall measurement system. The integration of the regulatory compliance measures into the overall measurement system insures that nothing is overlooked and that the compliance becomes an everyday part of each employee's job.

Conclusion

A well-designed, properly managed maintenance management system will help to insure regulatory compliance for most companies. However, companies that neglect "Best Practices" in the maintenance management discipline will, most likely, eventually incur some type of regulatory violation. As the book progresses, the interrelationships between the compliance standards and the maintenance Best Practices areas will be continually highlighted.

Preventive Maintenance

Preventive maintenance (PM) programs concentrate on the facility and equipment inspections as well as proactive correction of any defective or deteriorating conditions that are observed.

Types of Regulations

Consider the various regulations in the order we would use if we were performing a PM inspection.

General Housekeeping
Aisle Ways
Guardrails
Fixed Ladders
Building Structures
Building Support Equipment

General Housekeeping

Regulations state the following about general housekeeping:

> All places of employment shall be kept clean to the extent that the nature of the work allows.
>
> The floor of every workroom shall be maintained, so far as practicable, in a dry condition. Where wet processes are used, drainage shall be maintained and false floors, platforms, mats, or other dry standing places shall be provided, where practical, or appropriate waterproof footgear shall be provided.
>
> All sweepings, solid or liquid wastes, refuse, and garbage shall be removed in such a manner as to avoid creating a menace to

health and as often as necessary or appropriate to maintain the place of employment in a sanitary condition.

Vermin control. Every enclosed workplace shall be so constructed, equipped, and maintained, so far as reasonably practicable, as to prevent the entrance or harborage of rodents, insects, and other vermin. A continuing and effective extermination program shall be instituted where their presence is detected.

While these regulations extend beyond just the maintenance department, it is usually the responsibility of the maintenance organization to clean their own work or shop areas. Maintenance shop cleanliness is implied in the above standard.

In FDA regulated plants, the following are required guidelines:

Plant and grounds.

(a) Grounds. The grounds about a food plant under the control of the operator shall be kept in a condition that will protect against the contamination of food. The methods for adequate maintenance of grounds include, but are not limited to:

(1) Properly storing equipment, removing litter and waste, and cutting weeds or grass within the immediate vicinity of the plant buildings or structures that may constitute an attractant, breeding place, or harborage for pests.

(2) Maintaining roads, yards, and parking lots so that they do not constitute a source of contamination in areas where food is exposed.

(3) Adequately draining areas that may contribute contamination to food by seepage, foot-borne filth, or providing a breeding place for pests.

(4) Operating systems for waste treatment and disposal in an adequate manner so that they do not constitute a source of contamination in areas where food is exposed.

If the plant grounds are bordered by grounds not under the operator' control and not maintained in the manner described in paragraph (a) (1) through (3) of this section, care shall be exercised in the plant by inspection, extermination, or other means to exclude pests, dirt, and filth that may be a source of food contamination.

These regulations are more specific, leading to the cleaner environment that is required in a food-producing plant.

In addition, there are specific regulations concerning food consumption in a plant. These include:

Application.

This paragraph shall apply only where employees are permitted to consume food or beverages, or both, on the premises.

Eating and drinking areas.

No employee shall be allowed to consume food or beverages in a toilet room nor in any area exposed to a toxic material.

Waste disposal containers.

Receptacles constructed of smooth, corrosion resistant, easily cleanable, or disposable materials, shall be provided and used for the disposal of waste food. The number, size, and location of such receptacles shall encourage their use and not result in overfilling. They shall be emptied not less frequently than once each working day, unless unused, and shall be maintained in a clean and sanitary condition. Receptacles shall be provided with a solid tight-fitting cover unless sanitary conditions can be maintained without use of a cover.

Sanitary storage.

No food or beverages shall be stored in toilet rooms or in an area exposed to a toxic material.

Nonpotable water.

Outlets for nonpotable water, such as water for industrial or firefighting purposes, shall be posted or otherwise marked in a manner that will indicate clearly that the water is unsafe and is not to be used for drinking, washing of the person, cooking, washing of food, washing of cooking or eating utensils, washing of food preparation or processing premises, or personal service rooms, or for washing clothes.

The regulations clearly place an emphasis on providing a safe, clean workplace, where neither the employees or the customer for the products will be subject to contracting disease due to unclean practices.

Aisle Ways

'Sub-part D (Walking – Working Surfaces) includes aisle ways. Aisle ways must be clearly marked with lines of 2" to 6" in width and the aisle must be at least 3 feet wider than the largest equipment utilizing the aisle way or a minimum of 4 feet."

Why is this particular point of concern to a maintenance organization? It is that maintenance shops have requirements for aisles. What type of equipment will be brought in and out of the shop? Are the aisles into and out of the shop kept clear of equipment, so as not to obstruct the aisle? In addition, when maintenance is performed on equipment, are the aisles around the equipment kept free and clear? Or are they marked off, so no one tries to use them during the repair? Failure to do so could result in a violation and subsequent penalties.

Guard Rails

1910.23 states the following:

Standard railing shall consist of top rail, intermediate rail, and posts, and shall have a vertical height of 42 inches nominal from upper surface of top rail to floor, platform, runway, or ramp level. The top rail shall be smooth-surfaced throughout the length of the railing. The intermediate rail shall be approximately halfway between the top rail and the floor, platform, runway, or ramp. The ends of the rails shall not overhang the terminal posts except where such overhang does not constitute a projection hazard".

Posts shall be not more than eight (8) feet apart; they are to be permanent and substantial, smooth, and free from protruding nails, bolts, and splinters. If made of pipe, the post shall be one and one-fourth (1 1/4) inches inside diameter, or larger.

In an industrial setting, handrails are often damaged by material handling equipment, maintenance activities, or other small incidents. Damage to top and intermediate rails is quite common. However, when work orders are written to correct the damage, low priority is given to the work and they are often purged when companies try to reduce the size of their maintenance backlog. How can a company be said to be in compliance if the top or intermediate rail is not "smoothed-surfaced"? It is quite evident when walking through many plants that increased attention needs to be given to this area.

The standard continues with comments about toe boards. "A standard toe board shall be 4 inches nominal in vertical height from its top edge to the level of the floor, platform, runway, or ramp. It shall be securely fastened in place and with not more than 1/4-inch clearance above floor level. It may be made of any substantial material either solid or with openings not over 1 inch in greatest dimension." The basic reason for this regulation is to prevent items from falling off the walkway and injuring someone below. This generally relates to housekeeping. If the walkway is kept clear, then the toe boards can easily comply. However, in many plants, it is evident that items do get stored on walkways. Many of these items are over 4" in height. This is the reason for the following to be added to the regulation:

"Where material is piled to such height that a standard toe board does not provide protection, paneling from floor to intermediate rail, or to top rail shall be provided."

This provision allows for storage of items over 4" tall, given the additional required panel. When writing PM standards for a given area, you must check housekeeping. This check insures that unnecessary items are not stored in inappropriate locations that may violate this particular standard. Appropriate time should be allocated during the PM for the assigned maintenance technicians to give attention to proper housekeeping regulations.

One final item, which is generally more of a design/construction issue is the proper placement of handrails. The regulations state "All handrails and railings shall be provided with a clearance of not less than 3 inches between the handrail or railing and any other object.". Adherence to this regulation during the construction of handrails will prevent the typical "caught between" accident.

Fixed Ladders

Fixed ladders are ladders that are attached to a building or support structure. They may be embedded into concrete or attached to a building, silo, stack or other structure. There are specific requirements for these ladders. While it may appear this is design related, the maintenance department is often called upon to repair these ladders. The following regulations must be adhered to while effecting these repairs.

There are specific size requirements and structural dimensions for fixed ladders. Some of the requirements are:

All rungs shall have a minimum diameter of three-fourths inch for metal ladders

Acess ladders formed by individual metal rungs imbedded in concrete, which serve as access to pits and to other areas under floors, are frequently located in an atmosphere that causes corrosion and rusting. To increase rung life in such atmosphere, individual metal rungs shall have a minimum diameter of 1 inch or shall be painted or otherwise treated to resist corrosion and rusting.

All rungs will have a minimum diameter of 11/8 inches for wood ladders.

The distance between rungs, cleats, and steps shall not exceed 12 inches and shall be uniform throughout the length of the ladder.

The minimum clear length of rungs or cleats (actual width) shall be 16 inches.

Rungs, cleats, and steps shall be free of splinters, sharp edges, burrs, or projections which may be a hazard.

Metal ladders and appurtenances shall be painted or otherwise treated to resist corrosion and rusting when location demands.

Wood ladders, when used under conditions where decay may occur, shall be treated with a nonirritating preservative, and the details shall be such as to prevent or minimize the accumulation of water on wood parts.

Cages shall extend a minimum of 42 inches above the top of landing, unless other acceptable protection is provided.

Cages shall extend down the ladder to a point not less than 7 feet nor more than 8 feet above the base of the ladder, with bottom flared not less than 4 inches, or portion of cage opposite ladder shall be carried to the base.

Cages shall not extend less than 27 nor more than 28 inches from the centerline of the rungs of the ladder. Cage shall not be less than 27 inches in width. The inside shall be clear of projections. Vertical bars shall be located at a maximum spacing of 40 degrees around the circumference of the cage; this will give a maximum spacing of approximately 9 1/2 inches, center to center.

Ladder wells shall have a clear width of at least 15 inches measured each way from the centerline of the ladder. Smooth-walled wells shall be a minimum of 27 inches from the centerline of rungs to the well wall on the climbing side of the ladder. Where other obstructions on the climbing side of the ladder exist, there shall be a minimum of 30 inches from the centerline of the rungs.

The regulations continue with a section on maintenance which says:

"Maintenance." All ladders shall be maintained in a safe condition. All ladders shall be inspected regularly, with the intervals between inspections being determined by use and exposure.

The problem with this regulation is the term "inspected regularly". In a letter of clarification on a question about scaffolding, OSHA says:

What constitutes "periodic"

These standards do not specify how often a scaffold must be inspected to meet the "periodic" requirement. The frequency

> will depend on factors such as the type of scaffold, site and weather conditions, intensity of use, age of the equipment, and how often sections or components are added, removed or changed. These kinds of factors will determine how quickly or slowly safety related faults, loose connections, degradation and other defects can be expected to develop.
>
> "Periodic" means frequently enough so that, in light of these factors and the amount of time expected for their detrimental effects to occur, there is a good likelihood that problems will be found before they pose a hazard to employees.

It would be safe to assume that the definition on periodic and regular would have a similar intent. Therefore, establishing inspection frequencies depends on the plant and the area of the plant where the fixed ladder is located. It will be up to the individual setting up the inspection program to determine the "regularity" of the inspection.

As with previous regulations, a good, detailed, and documented Preventive Maintenance inspection will help a company maintain compliance with the regulations.

Building Structures

This section of the regulations (1910 Subpart E) deals with building access, building systems, and fire systems. The regulations in this section are designed to prevent injury to employees and property in the event of an accident. For example:

> Every required exit, way of approach thereto, and way of travel from the exit into the street or open space, shall be continuously maintained free of all obstructions or impediments to full instant use in the case of fire or other emergency.

During any safety or maintenance inspection, all stairwells must be open and free from any obstruction. This requirement eliminates anyone storing anything in the stairwell, even on a temporary basis. Further inspection and maintenance requirements are implied in the regulation:

> Every automatic sprinkler system, fire detection and alarm system, exit lighting, fire door, and other item of equipment, where

provided, shall be continuously in proper operating condition.

This regulation doesn't require a periodic inspection, but rather "continuously in proper operating condition". Therefore, it requires more frequent inspection and maintenance of the related equipment. The regulations are somewhat more vague in the following statement:

> Doors, stairs, ramps, passages, signs, and all other components of means of egress shall be of substantial, reliable construction and shall be built or installed in a workmanlike manner.

However, if the equipment is to be installed in a more workmanlike manner, it is only reasonable that they would be maintained in such a manner. This expectation necessitates the timely repair of any inspection item found to be deficient.

Building Exits

In view of many incidents of fire that have resulted in loss of life due to exits not being clearly marked or in some cases locked, the regulations statc the following:

> Any device or alarm installed to restrict the improper use of an exit shall be so designed and installed that it cannot, even in cases of failure, impede or prevent emergency use of such exit.
>
> Every exit sign shall be distinctive in color and shall provide contrast with decorations, interior finish, or other signs.
>
> A sign reading "Exit", or similar designation, with an arrow indicating the directions, shall be placed in every location where the direction of travel to reach the nearest exit is not immediately apparent.
>
> Every exit sign shall be suitably illuminated by a reliable light source giving a value of not less than 5 foot-candles on the illuminated surface. Artificial lights giving illumination to exit signs other than the internally illuminated types shall have screens, discs, or lenses of not less than 25 square inches area made of translucent material to show red or other specified designating color on the side of the approach.

> Each internally illuminated exit sign shall be provided in all occupancies where reduction of normal illumination is permitted.
>
> Every exit sign shall have the word "Exit" in plainly legible letters not less than 6 inches high, with the principal strokes of letters not less than three-fourths-inch wide.

Many maintenance requirements are either directly stated or implied in this set of standards. In most plants, the exits signs are specified and purchased, but the maintenance department installs, repairs, or changes the signs. The inspection of the signs for proper placement, lighting and coloration is important. Exit signs provide another example of how a company can be assisted with regulatory compliance through proper preventive maintenance, corrective maintenance, and work planning and scheduling. The areas of the maintenance program which impact these activities include preventive maintenance, inventory and purchasing, work flow, and CMMS usage.

Building Support Equipment

Roof Guarding and Building Repair Equipment

While the roof construction and guarding is more of a design than a maintenance consideration, there are still maintenance issues to address. If, for some reason, the roof guarding becomes damaged, needs replacement, or is removed during an installation or maintenance activity, the repair or replacement must be set to the OSHA standards to avoid a violation. Note some of the following requirements:

> When servicing equipment on roofs, particularly equipment used for cleaning or repairing the building, the fasteners used must be self-locking or otherwise designed to resist vibration and the subsequent loosening of the fasteners.
>
> All hoisting and repair equipment is to be serviced at set frequencies and the inspections are to be recorded and stored for reference by the appropriate personnel. The inspections are to be timed according to the manufacturer's recommendation, but not less than once per year. This detailed in-spection is to insure safe operation of the equipment.

A maintenance inspection of the repair platform equipment is required every 30 days, at the minimum. The repair platform is also to be inspected if it is used within the 30 day inspection window. As with the fixed inspections, documentation is to be kept on the date of the inspection, who performed the inspection, and specifically which repair platform was inspected. The inspection is to be performed by an individual who is deemed to be competent. The documentation would be tracked by the preventive maintenance program. Some general inspection areas to be detailed on the PM inspection include:

broken; worn, and damaged parts

switch contacts, brushes, and short flexible conductors of electrical devices

components of the electrical service system and traveling cables

gears, shafts, bearings, brakes and hoisting drums

The wire ropes used for support or hoisting are a specific part in the inspection. Before every use, they are to be inspected by a competent person. If there is any event that may cause damage to the wire ropes, they are to be re-inspected before further usage. The standards call for checking the following conditions during the inspection:

Broken wires exceeding three wires in one strand or six wires in one rope lay;

Distortion of rope structure such as would result from crushing or kinking;

Evidence of heat damage;

Evidence of rope deterioration from corrosion;

A broken wire within 18 inches (460.8 mm) of the end attachments;

Noticeable rusting and pitting;

Evidence of core failure (a lengthening of rope lay, protrusion of the rope core and a reduction in rope diameter suggests core failure); or

More than one valley break (broken wire).

Outer wire wear exceeds one-third of the original outer wire diameter.

Any other condition which the competent person determines has significantly affected the integrity of the rope.

Again, the reference is made to a competent person, which signifies training or knowledge that not all employees will possess, and a level of skill that will be required of the one completing the work. The documentation requirements require the inspection to be documented and the inspector to be properly identified.

In addition to having proper maintenance training, operators are also requiree to have proper training for inspection and operation of the building repair equipment. This training should include:

Recognition of, and preventive measures for, the safety hazards associated with their individual work tasks.

General recognition and prevention of safety hazards associated with the use of working platforms, including the provisions in the section relating to the particular working platform to be operated.

Emergency action plan procedures

Work procedures

Personal fall arrest system inspection, care, use and system performance

The training for the operators is not to be just on-the-job training. The regulations require written work instructions or pictorials. The manual for the operation and maintenance of the platform system supplied by the manufacturer may be used for the training. The standards also specify the training is to be a combination of both classroom and on-the-job training. The training can be obtained from the manufacturer (preferred) or it can be developed internally.

In addition the following requirements for tracking the training are detailed:

Upon completion of the training program, the employee should be able to demonstrate competency in operating the equipment

> safely. The employer, as necessary, should provide supplemental training of the employee, if the equipment used or other working conditions should change.
>
> The training must be certified by the company and the training record must contain the following information:
>
> > the identity of the person trained
> >
> > the signature of the employer or the person who conducted the training
> >
> > the date that training was completed.
>
> The training certification record must be kept on file until the employee leaves the company. The certification and training records must be produced at the request of an OSHA inspector.

These requirements clearly highlight that on-the-job training is insufficient to comply with the standards. The proper tracking of the training is essential, since this will be one of the first documents the inspector will require if an accident occurs.

Employee Personal Protective Equipment

Sometimes when performing maintenance activities, it is necessary for the employees to work in a hazardous environment. On these occasions, they are required to wear Personal Protective Equipment (PPE). Two common examples that are covered by regulatory agencies are respirators and hearing protectors. The general regulations for PPE include these statements:

> Protective equipment, including personal protective equipment for eyes, face, head, and extremities, protective clothing, respiratory devices, and protective shields and barriers, shall be provided, used, and maintained in a sanitary and reliable condition wherever it is necessary by reason of hazards of processes or environment, chemical hazards, radiological hazards, or mechanical irritants encountered in a manner capable of causing injury or impairment in the function of any part of the body through absorption, inhalation or physical contact.

Employee-owned equipment. Where employees provide their own protective equipment, the employer shall be responsible to assure its adequacy, including proper maintenance, and sanitation of such equipment.

Design. All personal protective equipment shall be of safe design and construction for the work to be performed.

Respirators

The primary goal of the regulations governing respirators is to prevent the open release of any contaminant. However, in cases where they can not be eliminated or contained, the following applies:

In the control of those occupational diseases caused by breathing air contaminated with harmful dusts, fogs, fumes, mists, gases, smokes, sprays, or vapors, the primary objective shall be to prevent atmospheric contamination. This shall be accomplished as far as feasible by accepted engineering control measures (for example, enclosure or confinement of the operation, general and local ventilation, and substitution of less toxic materials). When effective engineering controls are not feasible, or while they are being instituted, appropriate respirators shall be used pursuant to this section.

When respirators are required, it is the responsibility of the employer to provide the respirators.

Respirators shall be provided by the employer when such equipment is necessary to protect the health of the employee. The employer shall provide the respirators which are applicable and suitable for the purpose intended. The employer shall be re-sponsible for the establishment and maintenance of a respiratory protection program which shall include the requirements outlined in this section.

The maintenance of the respirators is important, especially in a maintenance department. Since the maintenance activities may require a variety of motions, the respirators will be subject to more rapid wear and the possibility of damage. The following regulations will apply:

Maintenance and care of respirators

This paragraph requires the employer to provide for the cleaning and disinfecting, storage, inspection, and repair of respirators used by employees.

Cleaning and disinfecting.

The employer shall provide each respirator user with a respirator that is clean, sanitary, and in good working order. The employer shall ensure that respirators are cleaned and disinfected using the procedures in Appendix B-2 of this section, or procedures recommended by the respirator manufacturer, provided that such procedures are of equivalent effectiveness. The respirators shall be cleaned and disinfected at the following intervals:

Respirators issued for the exclusive use of an employee shall be cleaned and disinfected as often as necessary to be maintained in a sanitary condition;

Respirators issued to more than one employee shall be cleaned and disinfected before being worn by different individuals;

Respirators maintained for emergency use shall be cleaned and disinfected after each use; and

Respirators used in fit testing and training shall be cleaned and disinfected after each use.

Storage.

The employer shall ensure that respirators are stored as follows:

All respirators shall be stored to protect them from damage, contamination, dust, sunlight, extreme temperatures, excessive moisture, and damaging chemicals, and they shall be packed or stored to prevent deformation of the facepiece and exhalation valve.

In addition to the requirements of paragraph (h)(2)(i) of this section, emergency respirators shall be:

Kept accessible to the work area;

Stored in compartments or in covers that are clearly marked as containing emergency respirators; and

Stored in accordance with any applicable manufacturer instructions.

Inspection.

The employer shall ensure that respirators are inspected as follows:

All respirators used in routine situations shall be inspected before each use and during cleaning;

All respirators maintained for use in emergency situations shall be inspected at least monthly and in accordance with the manufacturer's recommendations, and shall be checked for proper function before and after each use; and

Emergency escape-only respirators shall be inspected before being carried into the workplace for use.

The employer shall ensure that respirator inspections include the following:

A check of respirator function, tightness of connections, and the condition of the various parts including, but not limited to, the facepiece, head straps, valves, connecting tube, and cartridges, canisters or filters; and

A check of elastometric parts for pliability and signs of deterioration.

The above regulations may only seem common sense, however, consider the individual tracking and inspection of each respirator. Even though the employer may make employees responsible for their personal respirators, it is still the employer who is legally responsible if someone is injured using a defective or worn respirator.

Beyond respirators are self-contained breathing apparatus that are sometimes utilized where respirators are unsafe, such as in hazardous environments with chemical vapors. There are additional requirements for these styles of equipment.

In addition to the requirements of the above section, self-contained breathing apparatus shall be inspected monthly. Air and oxygen cylinders shall be maintained in a fully charged state and shall be recharged when the pressure falls to 90% of the manufacturer's recommended pressure level. The employer shall determine that the regulator and warning devices function properly.

In some remote locations, respirators may not be assigned to individual employees, but may be kept in a locker or suitable cabinet for emergency usage only. In these cases, the following guidelines apply:

For respirators maintained for emergency use, the employer shall:

(A) Certify the respirator by documenting the date the inspection was performed, the name (or signature) of the person who made the inspection, the findings, required remedial action, and a serial number or other means of identify ing the inspected respirator; and
(B) Provide this information on a tag or label that is attached to the storage compartment for the respirator, is kept with the respirator, or is included in inspection reports stored as paper or electronic files. This information shall be maintained until replaced following a subsequent certification.

Repairs.

The previous section highlighted the need for some type of inspection recordkeeping system, whether a manual or computerized system. As this next section indicates, not only inspection records, but also repair records must be maintained.

The employer shall ensure that respirators that fail an inspection or are otherwise found to be defective are removed from service, and are discarded or repaired or adjusted in accordance with the following procedures:

Repairs or adjustments to respirators are to be made only by persons appropriately trained to perform such operations and shall use only the respirator manufacturer's NIOSH-approved parts designed for the respirator;

Repairs shall be made according to the manufacturer's recommendations and specifications for the type and extent of repairs to be performed; and

Reducing and admission valves, regulators, and alarms shall be adjusted or repaired only by the manufacturer or a technician trained by the manufacturer.

Training and information.

Further data that is required to be tracked include the employee training in the safe use of respirators and other safety equipment. While these records may be kept manually, an organization of any relative size would do well to have a computerized database for this type of data tracking.

This paragraph requires the employer to provide effective training to employees who are required to use respirators. The training must be comprehensive, understandable, and recur annually, and more often if necessary. This paragraph also requires the employer to provide the basic information on respirators in Appendix D of this section to employees who wear respirators when not required by this section or by the employer to do so.

The employer shall ensure that each employee can demonstrate knowledge of at least the following:

Why the respirator is necessary and how improper fit, usage, or maintenance can compromise the protective effect of the respirator;

What the limitations and capabilities of the respirator are;

How to use the respirator effectively in emergency situations, including situations in which the respirator malfunctions;

How to inspect, put on and remove, use, and check the seals of the respirator;

What the procedures are for maintenance and storage of the respirator;

How to recognize medical signs and symptoms that may limit or prevent the effective use of respirators; and

The general requirements of this section.

The training shall be conducted in a manner that is understandable to the employee.

The employer shall provide the training prior to requiring the employee to use a respirator in the workplace.

Retraining shall be administered annually, and when the following situations occur:

Changes in the workplace or the type of respirator render previous training obsolete;

Inadequacies in the employee's knowledge or use of the respirator indicate that the employee has not retained the requisite understanding or skill; or

Any other situation arises in which retraining appears necessary to ensure safe respirator use.

The basic advisory information on respirators, as presented in Appendix D of this section, shall be provided by the employer in any written or oral format, to employees who wear respirators when such use is not required by this section or by the employer.

Recordkeeping.

This section requires the employer to establish and retain written information regarding medical evaluations, fit testing, and the respirator program. This information will facilitate employee involvement in the respirator program, assist the employer in auditing the adequacy of the program, and provide a record for compliance determinations by OSHA.

If respirator usage is required in the workplace, it can be seen that there is a considerable level of resources required to insure the company is in compliance and does not incur any fines when the respirator program is audited.

Noise Prevention

The second major area for PPE consideration is hearing protection, which is required for all employees who are going to be exposed to a noise level of 85 decibels for a period of 8 hours. This protection is to be provided to the employees at no cost to the employees. Also, the employer is required to replace the hearing protection as necessary to insure the ongoing protection of the employees. The employers are required to insure that hearing protection is worn as necessary.

If employees choose to supply their own hearing protection equipment, it is the responsibility of the employer to insure that the equipment is adequate. The employer must also insure that the employee-owned equipment is properly maintained and sanitary.

The employer is required to provide a training program for the employees in the noise protection program and must insure the employee's participation in the training program. The employer is also required to repeat the training program annually for each employee. The annual training update must include any changes in protective equipment and work processes.

The training program should include:

- The effects of noise levels on hearing
- The purpose of hearing protectors
- The advantages and disadvantages of the various types of hearing protectors
- When to use each type of hearing protection
- How to care for the hearing protection

The employer must then provide the training materials and information relating to the noise protection standard to all employees.

This area is another example of staff resources that are required to insure the employer does not fall into a regulatory violation and incur a fine.

Specialized Equipment

This section contains some preventive maintenance guidelines for specific types of equipment that are common in many plants.

Cranes

The following regulations apply to various types of cranes. However, the inspection and maintenance of all cranes is considered a priority with most OSHA inspectors. It would be good to consider the PM requirements for cranes.

The first regulation applies to outdoor cranes.

Wind indicators and rail clamps.

Outdoor storage bridges shall be provided with automatic rail clamps. A wind-indicating device shall be provided which will give a visible or audible alarm to the bridge operator at a predetermined wind velocity. If the clamps act on the rail heads, any beads or weld flash on the rail heads shall be ground off.

This regulation insures that a crane operator will not remain on the crane when unsafe wind conditions arise. There are a set of clamps that fasten the crane to the rail, locking it in place, and insuring it will not blow over during high winds. However, if the clamps are not adjusted correctly or if they are applied while the crane is in motion, they may slide against the rail. Any damage to the rail, must be repaired before the crane is put back into use.

Access to crane.

Access to the cab and/or bridge walkway shall be by a conveniently placed fixed ladder, stairs, or platform requiring no step over any gap exceeding 12 inches.

This regulation insures it is safe for the operator to access the cab of the crane. All ladders and walkways must meet the standards discussed earlier in this chapter.

The following regulations deal with safety mechanisms used for cranes. The bumpers are designed to prevent the crane from running off the end of the track and the rail sweeps are designed to keep any object from derailing the crane. While it may seem unlikely for these events to occur, they actually have on numerous occasions.

Bridge bumpers.

A crane shall be provided with bumpers or other automatic means providing equivalent effect, unless the crane travels at a slow rate of speed and has a faster deceleration rate due to the use of sleeve bearings, or is not operated near the ends of bridge and trolley travel, or is restricted to a limited distance by the nature of the crane operation and there is no hazard of striking any object in this limited distance, or is used in similar operating conditions. The bumpers shall be capable of stopping the crane (not including the lifted load) at an average rate of deceleration not to exceed 3 ft/s/s when traveling in either direction at 20 percent of the rated load speed.

The bumpers shall have sufficient energy absorbing capacity to stop the crane when traveling at a speed of at least 40 percent of rated load speed.

The bumper shall be so mounted that there is no direct shear on bolts.

Rail sweeps.

Bridge trucks shall be equipped with sweeps which extend below the top of the rail and project in front of the truck wheels.

Guards

These regulations are essential inspection points for preventive maintenance checklists.

Hoisting ropes.

If hoisting ropes run near enough to other parts to make fouling or chafing possible, guards shall be installed to prevent this condition.

A guard shall be provided to prevent contact between bridge conductors and hoisting ropes if they could come into contact.

Guards for moving parts.

Exposed moving parts such as gears, set screws, projecting keys, chains, chain sprockets, and reciprocating components which might constitute a hazard under normal operating conditions shall be guarded.

Guards shall be securely fastened.
Each guard shall be capable of supporting without permanent distortion the weight of a 200-pound person unless the guard is located where it is impossible for a person to step on it.

While it would be unusual for someone to be on the crane while it is in operation, maintenance personnel may occasionally do so while troubleshooting or making observation inspections. Hence, the regulation on guards has high importance.

Brakes

There may be three or more braking systems on a crane. These include a bridge brake, a hoist brake, and a trolley brake. The following regulations apply:

Holding brakes on hoists shall be applied automatically when power is removed.

Where necessary, holding brakes shall be provided with adjustment means to compensate for wear.

The wearing surface of all holding-brake drums or discs shall be smooth

Each independent hoisting unit of a crane handling hot metal and having power control braking means shall be equipped with at least two holding brakes.

The wearing surface of all brake drums or discs shall be smooth

Electric Equipment

The electrical equipment on the crane must meet certain standards. The PM inspection should always include and inspection of the electrical systems to insure compliance.

General.
Wiring and equipment shall comply with subpart S of this part.

The control circuit voltage shall not exceed 600 volts for a.c. or d.c. current.

The voltage at pendant push-buttons shall not exceed 150 volts for a.c. and 300 volts for d.c.

Equipment.
Electrical equipment shall be so located or enclosed that live parts will not be exposed to accidental contact under normal operating conditions.

Electric equipment shall be protected from dirt, grease, oil, and moisture.

Guards for live parts shall be substantial and so located that they cannot be accidentally deformed so as to make contact with the live parts.

Hoisting Systems

One of the most important parts of the crane is the hoisting system. The criticality is due to the fact that the hoist will be performing tasks in conjunction with other equipment and possibly other employees. Thus, the hoisting mechanism is given the utmost scrutiny in any PM inspection. The regulations specify:

The hoisting motion of all electric traveling cranes shall be provided with an overtravel limit switch in the hoisting direction

Sheaves carrying ropes which can be momentarily unloaded shall be provided with close-fitting guards or other suitable devices to guide the rope back into the groove when the load is applied again.

The sheaves in the bottom block shall be equipped with close-fitting guards that will prevent ropes from becoming fouled when the block is lying on the ground with ropes loose.

Pockets and flanges of sheaves used with hoist chains shall be of such dimensions that the chain does not catch or bind during operation.

All running sheaves shall be equipped with means for lubrication. Permanently lubricated, sealed and/or shielded bearings meet this requirement.

No less than two wraps of rope shall remain on the drum when the hook is in its extreme low position.

Rope ends shall be anchored by a clamp securely attached to the drum, or by a socket arrangement approved by the crane or rope manufacturer.

Replacement rope shall be the same size, grade, and construction as the original rope furnished by the crane manufacturer, unless otherwise recommended by a wire rope manufacturer due to actual working condition requirements.

Equalizers.
If a load is supported by more than one part of rope, the tension in the parts shall be equalized.

Overall, due to the potential for damage by the crane, the OSHA regulations are detailed about the inspections and the frequency of the inspections. The following excepts from the regulations address these areas:

Inspection procedure for cranes in regular service is divided into two general classifications based upon the intervals at which inspection should be performed. The intervals in turn are dependent upon the nature of the critical components of the crane and the degree of their exposure to wear, deterioration, or malfunction. The two general classifications are herein designated as “frequent” and “periodic” with respective intervals between inspections as defined below:

Frequent inspection - *Daily to monthly intervals.*

Periodic inspection - *1 to 12-month intervals.*

Frequent inspection.

The following items shall be inspected for defects at intervals as defined in paragraph (j)(1)(ii) of this section or as specifically indicated, including observation during operation for any defects which might appear between regular inspections. All deficiencies such as listed shall be carefully examined and determination made as to whether they constitute a safety hazard:

All functional operating mechanisms for maladjustment interfering with proper operation. Daily.

Deterioration or leakage in lines, tanks, valves, drain pumps, and other parts of air or hydraulic systems. Daily.

Hooks with deformation or cracks. Visual inspection daily; monthly inspection with a certification record which includes the date of inspection, the signature of the person who performed the inspection and the serial number, or other identifier, of the hook inspected. For hooks with cracks or having more than 15 percent in excess of normal throat opening or more than 10 twist from the plane of the unbent hook refer to paragraph (l)(3)(iii)(a) of this section.

Hoist chains, including end connections, for excessive wear, twist, distorted links interfering with proper function, or stretch beyond manufacturer's recommendations. Visual inspection daily; monthly inspection with a certification record which includes the date of inspection, the signature of the person who performed the inspection and an identifier of the chain which was inspected.

All functional operating mechanisms for excessive wear of components.

Rope reeving for noncompliance with manufacturer's recommendations.

Periodic inspection.

Complete inspections of the crane shall be performed at intervals as generally defined in paragraph (j)(1)(ii)(b) of this section, depending upon its activity, severity of service, and environment, or as specifically indicated below. These inspections shall include the requirements of paragraph (j)(2) of this section and in addition, the following items. Any deficiencies such as listed shall be carefully examined and determination made as to whether they constitute a safety hazard:

Deformed, cracked, or corroded members.

Loose bolts or rivets.

Cracked or worn sheaves and drums.

Worn, cracked or distorted parts such as pins, bearings, shafts, gears, rollers, locking and clamping devices.

Excessive wear on brake system parts, linings, pawls, and ratchets.

Load, wind, and other indicators over their full range, for any significant inaccuracies.

Gasoline, diesel, electric, or other power plants for improper performance or noncompliance with applicable safety requirements.

Excessive wear of chain drive sprockets and excessive chain stretch.

Electrical apparatus, for signs of pitting or any deterioration of controller contactors, limit switches and pushbutton stations.

Cranes not in regular use.

A crane which has been idle for a period of 1 month or more, but less than 6 months, shall be given an inspection conforming with requirements of paragraph (j)(2) of this section and paragraph (m)(2) of this section before placing in service.

A crane which has been idle for a period of over 6 months shall be given a complete inspection conforming with requirements of paragraphs (j)(2) and (3) of this section and paragraph (m)(2) of this section before placing in service.

Standby cranes shall be inspected at least semi-annually in accordance with requirements of paragraph (j)(2) of this section and paragraph (m)(2) of this section.

The trip setting of hoist limit switches shall be determined by tests with an empty hook traveling in increasing speeds up to the maximum speed. The actuating mechanism of the limit switch shall be located so that it will trip the switch, under all conditions, in sufficient time to prevent contact of the hook or hook block with any part of the trolley.

While those guidelines seem specific, they are still not sufficient. The regulations go on to state that:

A preventive maintenance program based on the crane manufacturer's recommendations shall be established.

Preventive Maintenance of Cranes

The specific PM program recommended by the manufacturer will vary somewhat, but in no circumstance should the PM program be less than the details provided in the standard. In fact, most manufacturers provide considerable detail for all preventive maintenance inspections and service. Some additional details for the inspection should include a safety shutdown procedure:

Before adjustments and repairs are started on a crane the following precautions shall be taken:

The crane to be repaired shall be run to a location where it will cause the least interference with other cranes and operations in the area.

All controllers shall be at the off position.

The main or emergency switch shall be open and locked in the open position.

Warning or "out of order" signs shall be placed on the crane, also on the floor beneath or on the hook where visible from the floor.

Where other cranes are in operation on the same runway, rail stops or other suitable means shall be provided to prevent interference with the idle crane.

After adjustments and repairs have been made the crane shall not be operated until all guards have been reinstalled, safety devices reactivated and maintenance equipment removed.

During the preventive maintenance service, the following should be considered:

Adjustments and repairs.

Any unsafe conditions disclosed by the inspection requirements of paragraph (j) of this section shall be corrected before operation of the crane is resumed. Adjustments and repairs shall be done only by designated personnel.

Adjustments shall be maintained to assure correct functioning of components. The following are examples:

All functional operating mechanisms.

Limit switches.

Control systems.

Brakes.

Power plants.

Repairs or replacements shall be provided promptly as needed for safe operation. The following are examples:

Crane hooks showing defects described in paragraph (j)(2)(iii) of this section shall be discarded. Repairs by welding or reshaping are not generally recommended. If such repairs are attempted they shall only be done under competent supervision and the hook shall be tested to the load requirements of paragraph (k)(2) of this section before further use.

Load attachment chains and rope slings showing defects described in paragraph (j)(2) (iv) and (v) of this section respectively.

All critical parts which are cracked, broken, bent, or excessively worn.

Pendant control stations shall be kept clean and function labels kept legible.

ROPE INSPECTIONS

Running ropes.

A thorough inspection of all ropes shall be made at least once a month and a certification record which includes the date of inspection, the signature of the person who performed the inspection and an identifier for the ropes which were inspected shall be kept on file where readily available to appointed personnel. Any deterioration, resulting in appreciable loss of original strength, shall be carefully observed and determination made as to whether further use of the rope would constitute a safety hazard. Some of the conditions that could result in an appreciable loss of strength are the following:

Reduction of rope diameter below nominal diameter due to loss of core support, internal or external corrosion, or wear of outside wires.

A number of broken outside wires and the degree of distribution or concentration of such broken wires.

Worn outside wires.

Corroded or broken wires at end connections.

Corroded, cracked, bent, worn, or improperly applied end connections.

Severe kinking, crushing, cutting, or unstranding.

Other ropes.

All rope which has been idle for a period of a month or more due to shutdown or storage of a crane on which it is installed shall be given a thorough inspection before it is used. This inspection shall be for all types of deterioration and shall be performed by an appointed person whose approval shall be required for further use of the rope. A certification record shall be available for inspection which includes the date of inspection, the signature of the person who performed the inspection and an identifier for the rope which was inspected.

Before starting to hoist the following conditions shall be noted:

Hoist rope shall not be kinked.

Multiple part lines shall not be twisted around each other.

The hook shall be brought over the load in such a manner as to prevent swinging.

During hoisting care shall be taken that:

There is no sudden acceleration or deceleration of the moving load.

The load does not contact any obstructions.

Cranes shall not be used for side pulls except when specifically authorized by a responsible person who has determined that the stability of the crane is not thereby endangered and that various parts of the crane will not be overstressed.

While any employee is on the load or hook, there shall be no hoisting, lowering, or traveling.

The employer shall require that the operator avoid carrying loads over people.

The operator shall test the brakes each time a load approaching the rated load is handled. The brakes shall be tested by raising the load a few inches and applying the brakes.

The load shall not be lowered below the point where less than two full wraps of rope remain on the hoisting drum.

When two or more cranes are used to lift a load one qualified responsible person shall be in charge of the operation. He shall analyze the operation and instruct all personnel involved in the proper positioning, rigging of the load, and the movements to be made.

The employer shall insure that the operator does not leave his position at the controls while the load is suspended.

When starting the bridge and when the load or hook approaches near or over personnel, the warning signal shall be sounded.

Hoist limit switch.

At the beginning of each operator's shift, the upper limit switch of each hoist shall be tried out under no load. Extreme care shall be exercised; the block shall be "inched" into the limit or run in at slow speed. If the switch does not operate properly, the appointed person shall be immediately notified.

The hoist limit switch which controls the upper limit of travel of the load block shall never be used as an operating control.

Additional inspection areas include:

Ladders.

The employer shall insure that hands are free from encumbrances while personnel are using ladders.

Articles which are too large to be carried in pockets or belts shall be lifted and lowered by hand line.

Cabs.

Necessary clothing and personal belongings shall be stored in such a manner as not to interfere with access or operation.

Tools, oil cans, waste, extra fuses, and other necessary articles shall be stored in the tool box, and shall not be permitted to lie loose in or about the cab.

Fire extinguishers. The employer shall insure that operators are familiar with the operation and care of fire extinguishers provided.

MACHINERY GUARDING

This section looks at the proper guarding of equipment. Guards are of particular concern to the maintenance department for two reasons:

1. The maintenance technicians generally have to remove and reinstall guards during PM inspections and service

2. When working around the equipment, the guards must protect not only the maintenance technicians, but also the operators in case of an equipment malfunction.

Some of the relevant regulations follow:

General requirements for machine guards.

Guards shall be affixed to the machine where possible and secured elsewhere if for any reason attachment to the machine is not possible. The guard shall be such that it does not offer an accident hazard in itself.

The point of operation of machines whose operation exposes an employee to injury shall be guarded. The guarding device shall be in conformity with any appropriate standards therefore, or, in the absence of applicable specific standards, shall be so designed and constructed as to prevent the operator from having any part of his body in the danger zone during the operating cycle.

Special handtools for placing and removing material shall be such as to permit easy handling of material without the operator placing a hand in the danger zone. Such tools shall not be in lieu of other guarding required by this section, but can only be used to supplement protection provided.

Barrels, containers, and drums.

Revolving drums, barrels, and containers shall be guarded by an enclosure which is interlocked with the drive mechanism, so that the barrel, drum, or container cannot revolve unless the guard enclosure is in place.

Exposure of blades.

When the periphery of the blades of a fan is less than seven (7) feet above the floor or working level, the blades shall be guarded. The guard shall have openings no larger than one-half (1/2) inch.

The following two examples for typical shop equipment illustrate how specific the regulations can be.:

Self-feed circular saws.

Feed rolls and saws shall be protected by a hood or guard to prevent the hands of the operator from coming in contact with the in-running rolls at any point. The guard shall be constructed of heavy material, preferably metal, and the bottom of the guard shall come down to within three-eighths inch of the plane formed by the bottom or working surfaces of the feed rolls. This distance (three-eighths inch) may be increased to three-fourths inch, provided the lead edge of the hood is extended to be not less than 5 1/2 inches in front of the nip point between the front roll and the work.

Bandsaws and band resaws.

All portions of the saw blade shall be enclosed or guarded, except for the working portion of the blade between the bottom of the guide rolls and the table. Bandsaw wheels shall be fully encased. The outside periphery of the enclosure shall be solid. The front and back of the band wheels shall be either enclosed by solid material or by wire mesh or perforated metal. Such mesh or perforated metal shall be not less than 0.037 inch (U.S. Gage No. 20), and the openings shall be not greater than three-eighths inch. Solid material used for this purpose shall be of an equivalent strength and firmness. The guard for the portion of the blade between the sliding guide and the upper-saw-wheel guard shall protect the saw blade at the front and outer side. This portion of the guard shall be self-adjusting to raise and lower with the guide. The upper-wheel guard shall be made to conform to the travel of the saw on the wheel.

While these are just two examples of guarding requirements, the specificity for much of the equipment is incredibly detailed. Thus, all guards should receive special consideration on all PM inspections and service.

OSHA issued a memorandum concerning the guarding of equipment in a food processing plant. The company wanted to omit the guards due to a potential contamination problem. The ultimate answer was the equipment had to be guarded. In part, the OSHA response said:

The Food and Drug Administration does not have any rules or regulations that prohibit the guarding of machinery or power transmission apparatus. Their position on this issue is that any guards must not cause any contamination of the food products.

The bottom line is all equipment must be properly guarded – no exceptions.

Machine Controls and Equipment

The following regulations are intended to provide guidelines for general electrical safety when operating or maintaining equipment.

> The frames and all exposed, noncurrent-carrying metal parts of portable electric woodworking machinery operated at more than 90 volts to ground shall be grounded and other portable motors driving electric tools which are held in the hand while being operated shall be grounded if they operate at more than 90 volts to ground. The ground shall be provided through use of a separate ground wire and polarized plug and receptacle.
>
> A mechanical or electrical power control shall be provided on each machine to make it possible for the operator to cut off the power from each machine without leaving his position at the point of operation.
>
> On applications where injury to the operator might result if motors were to restart after power failures, provision shall be made to prevent machines from automatically restarting upon restoration of power.
>
> On each machine operated by electric motors, positive means shall be provided for rendering such controls or devices inoperative while repairs or adjustments are being made to the machines they control.
>
> Anchoring fixed machinery. Machines designed for a fixed location shall be securely anchored to prevent walking or moving.

OSHA inspectors have regulations dealing with specific equipment systems. These systems take many of the general regulations and focus them on an entire equipment system, comprised on many of the already detailed subsystems.The following are some typical examples:

Ventilation Systems for Particle Blasting Environments.

This section looks specifically at one work environment. keep in mind: many regulations that only apply in one work environment can be interpreted to apply in other similar work environments. Therefore, work environments that are extremely dusty or have excessive airborne particles (blast furnace, coke plants, basic oxygen furnaces, concrete plants, foundaries, etc.) may have similar requirements.

The design, installation, and maintenance of ventilation systems for particle blasting operations are specified in standards that are incorporated into the regulations. The maintenance department has major responsibilities in insuring the company meets the regulations once the equipment is in service. For example, the regulations specify:

Dust leaks are to be repaired as soon as possible

Pressure drops at the exhaust ducts are to be baselined upon the initial startup of the equipment and then monitored. When an appreciable pressure drop is noted, the system is to be cleaned and returned to the baseline condition.

Abrasive separators and dust collectors are to be installed and properly maintained to prevent contamination of other areas. The periodic maintenance of this equipment will be crucial to compliance with the regulation.

In this part of the regulation, it is clearly seen that the preventive maintenance and work order systems are important to comply with the ventilation system maintenance and data tracking requirements.

Employees working in a particle-blasting environment have certain requirements for personal protective equipment. First, these personnel must wear clothing suitable to protect them from the blast particles. This clothing may include, but is not limited to, heavy canvas or leather gloves, aprons, and appropriate safety shoes.

Second, the employees must have the appropriate respiratory equipment and the employers must have a respiratory protection program in place. In addition, where the respirator does not provide proper protection for the face and eyes, the proper protective equipment shall be supplied by the employer.

Inspection and Maintenance of Woodworking Machinery

The inspections and services mentioned in this section focus on keeping the equipment and related tools in their optimum operating condition. Many of the regulations are designed to prevent the woodworking tool from jamming, causing the workpiece to move in an unpredictable manner, causing injury to the operator. For example:

> Dull, badly set, improperly filed, or improperly tensioned saws shall be immediately removed from service, before they begin to cause the material to stick, jam, or kick back when it is fed to the saw at normal speed. Saws to which gum has adhered on the sides shall be immediately cleaned.
>
> All knives and cutting heads of woodworking machines shall be kept sharp, properly adjusted, and firmly secured. Where two or more knives are used in one head, they shall be properly balanced.
>
> Bearings shall be kept free from lost motion and shall be well lubricated.
>
> Arbors of all circular saws shall be free from play.
>
> Sharpening or tensioning of saw blades or cutters shall be done only by persons of demonstrated skill in this kind of work.
>
> Emphasis is placed upon the importance of maintaining cleanliness around woodworking machinery, particularly as regards the effective functioning of guards and the prevention of fire hazards in switch enclosures, bearings, and motors.

All cracked saws shall be removed from service.

The practice of inserting wedges between the saw disk and the collar to form what is commonly known as a "wobble saw" shall not be permitted.

Push sticks or push blocks shall be provided at the work place in the several sizes and types suitable for the work to be done.

The knife blade of jointers shall be so installed and adjusted that it does not protrude more than one-eighth inch beyond the cylindrical body of the head. Push sticks or push blocks shall be provided at the work place in the several sizes and types suitable for the work to be done.

Whenever veneer slicers or rotary veneer-cutting machines have been shutdown for the purpose of inserting logs or to make adjustments, operators shall make sure that machine is clear and other workmen are not in a hazardous position before starting the machine.

These items may seem routine to an experienced technician, but what about a new technician or an apprentice? The PM should not just indicate to perform an inspection, but should provide details similar to the above.

Inspection and Maintenance of Grinding Equipment

Another common piece of equipment in most plants is a grinder. Among the types of grinders are pedestal, hand, and heavy equipment grinders. The inspection and safety items include the following:

Work rests.

On offhand grinding machines, work rests shall be used to support the work. They shall be of rigid construction and designed to be adjustable to compensate for wheel wear. Work rests shall be kept adjusted closely to the wheel with a maximum opening of one-eighth inch to prevent the work from being jammed between the wheel and the rest, which may cause wheel breakage. The work rest shall be securely clamped after each adjustment. The adjustment shall not be made with the wheel in motion.

Bench and floor stands.

The angular exposure of the grinding wheel periphery and sides for safety guards used on machines known as bench and floor stands should not exceed 90 deg. or one-fourth of the periphery. This exposure shall begin at a point not more than 65 deg. above the horizontal plane of the wheel spindle.

Exposure adjustment.

Safety guards of the types previously described, where the operator stands in front of the opening, shall be constructed so that the peripheral protecting member can be adjusted to the constantly decreasing diameter of the wheel. The maximum angular exposure above the horizontal plane of the wheel spindle as previously specified shall never be exceeded, and the distance between the wheel periphery and the adjustable tongue or the end of the peripheral member at the top shall never exceed one-fourth inch.

General requirements.

All abrasive wheels shall be mounted between flanges which shall not be less than one-third the diameter of the wheel.

Blotters.

Blotters (compressible washers) shall always be used between flanges and abrasive wheel surfaces to insure uniform distribution of flange pressure.

Inspection.

Immediately before mounting, all wheels shall be closely inspected and sounded by the user (ring test) to make sure they have not been damaged in transit, storage, or otherwise. The spindle speed of the machine shall be checked before mounting of the wheel to be certain that it does not exceed the maximum operating speed marked on the wheel. Wheels should be tapped gently with a light nonmetallic implement, such as the handle of a screwdriver for light wheels, or a wooden mallet for heavier wheels. If they sound cracked (dead), they shall not be used. This is known as the "Ring Test".

Wheels must be dry and free from sawdust when applying the ring test, otherwise the sound will be deadened. It should also be noted that organic bonded wheels do not emit the same clear metallic ring as do vitrified and silicate wheels.

"Tap" wheels about 45 deg. each side of the vertical centerline and about 1 or 2 inches from the periphery. Then rotate the wheel 45 deg. and repeat the test. A sound and undamaged wheel will give a clear metallic tone. If cracked, there will be a dead sound and not a clear "ring."

Arbor size.

Grinding wheels shall fit freely on the spindle and remain free under all grinding conditions. A controlled clearance between the wheel hole and the machine spindle (or wheel sleeves or adaptors) is essential to avoid excessive pressure from mounting and spindle expansion. To accomplish this, the machine spindle shall be made to nominal (standard) size plus zero minus .002 inch, and the wheel hole shall be made suitably oversize to assure safety clearance under the conditions of operating heat and pressure.

Surface condition.

All contact surfaces of wheels, blotters and flanges shall be flat and free of foreign matter.

After reviewing these regulations, it is clear that proper inspection and service of any type of grinders is an important safety issue. Thus, all grinding equipment must be maintained to insure a safe working environment. If they are not maintained properly, there will be considerable attention paid to them during an OSHA inspection.

General Power Transmission PM Inspection Guidelines

The following are some general guidelines for power transmission components. While they may seem to be brief, it is interesting to note that the guidelines set the timing for the PM inspection. They must take place at least once every two months.

Care of equipment.

General. All power-transmission equipment shall be inspected at intervals not exceeding 60 days and be kept in good working condition at all times.

Shafting.

Shafting shall be kept in alignment, free from rust and excess oil or grease.

Where explosives, explosive dusts, flammable vapors or flammable liquids exist, the hazard of static sparks from shafting shall be carefully considered.

Bearings.

Bearings shall be kept in alignment and properly adjusted.

Hangers.

Hangers shall be inspected to make certain that all supporting bolts and screws are tight and that supports of hanger boxes are adjusted properly.

Pulleys.

Pulleys shall be kept in proper alignment to prevent belts from running off.

Broken pulleys. Pulleys with cracks, or pieces broken out of rims, shall not be used

Care of belts.

Inspection shall be made of belts, lacings, and fasteners and such equipment kept in good repair.

Lubrication.

The regular oiler shall wear tight-fitting clothing. Machinery shall be oiled when not in motion, wherever possible.

This list provides insight into what must be included in the inspection of power transmission equipment and the details that are required.

Additional Equipment Considerations for FDA

While the information considered previously was derived from the OSHA standards, the following information is specific to FDA-regulated plants.

All plant equipment and utensils shall be so designed and of such material and workmanship as to be adequately cleanable, and shall be properly maintained.

An interesting perspective on the term "properly maintained." In whose opinion? Properly maintained can mean one thing to a nontechnical person and quite another to a technical person. Note that regulatory agencies favor the opinion of the technical person. They will use the opinion of an expert form the area of the plant they are examining. For example, a quality issue is handled by the quality department. Similarly, the maintenance department would determine what "properly maintained" signifies.

The design, construction, and use of equipmentand utensils shall preclude the adulteration of food with lubricants,fuel, metal fragments, contaminated water, or any other contaminants.

All equipment should be so installed and maintained as to facilitate the cleaning of the equipment and of all adjacent spaces.

Food-contact surfaces shall be corrosion-resistant when in contact with food. They shall be made of nontoxic materials and designed to withstand the environment of their intended use and the action of food, and, if applicable, cleaning compounds and sanitizing agents. Food-contact surfaces shall be maintained to protect food from being contaminated by any source, including unlawful indirect food additives.

Seams on food-contact surfaces shall be smoothly bonded or maintained so as to minimize accumulation of food particles, dirt, and organic matter and thus minimize the opportunity for growth of microorganisms.

Equipment that is in the manufacturing or food-handling area and that does not come into contact with food shall be so constructed that it can be kept in a clean condition.

Holding, conveying, and manufacturing systems, including gravimetric, pneumatic, closed, and automated systems, shall be of a design and construction that enables them to be maintained in an appropriate sanitary condition.

Each freezer and cold storage compartment used to store and hold food capable of supporting growth of microorganisms shall be fitted with an indicating thermometer, temperature-measuring device, or temperature-recording device so installed as to show the temperature accurately within the compartment, and should be fitted with an automatic control for regulating temperature or with an automatic alarm system to indicate a significant temperature change in a manual operation.

Instruments and controls used for measuring, regulating, or recording temperatures, pH, acidity, water activity, or other conditions that control or prevent the growth of undesirable microorganisms in food shall be accurate and adequately maintained, and adequate in number for their designated uses.

Compressed air or other gases mechanically introduced into food or used to clean food-contact surfaces or equipment shall be treated in such a way that food is not contaminated with unlawful indirect food additives.

This series of regulations provides some specific items that a PM program in an FDA regulated plant should contain. These same items will not receive the same level of attention in a non-FDA regulated plant.

Sanitary Operations

This section provides some additional considerations for any FDA-regulated plant. While cleanliness is a concern in any plant, in a FDA-regulated plant, it is of utmost importance due to the possibility of product contamination. Some of the specific guidelines include:

General maintenance.

Buildings, fixtures, and other physical facilities of the plant shall be maintained in a sanitary condition and shall be kept in repair sufficient to prevent food from becoming adulterated within the meaning of the act. Cleaning and sanitizing of utensils and equipment shall be conducted in a manner that protects against contamination of food, food-contact surfaces, or food-packaging materials.

In addition to outside contaminants, the possibility exists for chemical contamination from the materials used in sanitizing and cleaning the plant. The regulations continue:

Substances used in cleaning and sanitizing; storage of toxic materials.

(1) Cleaning compounds and sanitizing agents used in cleaning and sanitizing procedures shall be free from undesirable microor-

ganisms and shall be safe and adequate under the conditions of use. Compliance with this requirement may be verified by any effective means including purchase of these substances under a supplier's guarantee or certification, or examination of these substances for contamination. Only the following toxic materials may be used or stored in a plant where food is processed or exposed:

(i) Those required to maintain clean and sanitary conditions;
(ii) Those necessary for use in laboratory testing procedures;
(iii)Those necessary for plant and equipment maintenance and operation; and
(iv) Those necessary for use in the plant's operations.

Toxic cleaning compounds, sanitizing agents, and pesticide chemicals shall be identified, held, and stored in a manner that protects against contamination of food, food-contact surfaces, or food- packaging materials. All relevant regulations promulgated by other Federal, State, and local government agencies for the application, use, or holding of these products should be followed.

The actual line equipment may require sanitation procedures if it comes in direct contact with the food. These procedures increase the wear on the equipment and shorten its design life if the equipment is not specified correctly at the design phase. The additional water and chemicals will have an adverse effect on the equipment. However, the standards state:

Sanitation of food-contact surfaces.

All food-contact surfaces, including utensils and food-contact surfaces of equipment, shall be cleaned as frequently as necessary to protect against contamination of food.

(1) Food-contact surfaces used for manufacturing or holding low-moisture food shall be in a dry, sanitary condition at the time of use. When the surfaces are wet-cleaned, they shall, when necessary, be sanitized and thoroughly dried before subsequent use.

(2) In wet processing, when cleaning is necessary to protect against the introduction of microorganisms into food, all food-contact surfaces shall be cleaned and sanitized before use and after any interruption during which the food-contact surfaces

may have become contaminated. Where equipment and utensils are used in a continuous production operation, the utensils and food-contact surfaces of the equipment shall be cleaned and sanitized as necessary.

(3) Non-food-contact surfaces of equipment used in the operation of food plants should be cleaned as frequently as necessary to protect against contamination of food.

(4) Single-service articles (such as utensils intended for one-time use, paper cups, and paper towels) should be stored in appropriate containers and shall be handled, dispensed, used, and disposed of in a manner that protects against contamination of food or food-contact surfaces.

(5) Sanitizing agents shall be adequate and safe under conditions of use. Any facility, procedure, or machine is acceptable for cleaning and sanitizing equipment and utensils if it is established that the facility, procedure, or machine will routinely render equipment and utensils clean and provide adequate cleaning and sanitizing treatment.

Storage and handling of cleaned portable equipment and utensils.

Cleaned and sanitized portable equipment with food-contact surfaces and utensils should be stored in a location and manner that protects food-contact surfaces from contamination.

While cleaning and sanitation may seem to be straightforward processes, the FDA has some stringent recordkeeping requirements. For example,

> The FDA expects firms to have written procedures (SOP's) detailing the cleaning processes used for various pieces of equipment. If firms have one cleaning process for cleaning between different batches of the same product and use a different process for cleaning between product changes, we expect the written procedures to address these different scenario. Similarly, if firms have one process for removing water soluble residues and another process for non-water soluble residues, the written procedure should address both scenarios and make it clear when

a given procedure is to be followed. Bulk pharmaceutical firms may decide to dedicate certain equipment for certain chemical manufacturing process steps that produce tarry or gummy residues that are difficult to remove from the equipment. Fluid bed dryer bags are another example of equipment that is difficult to clean and is often dedicated to a specific product. Any residues from the cleaning process itself (detergents, solvents, etc.) also have to be removed from the equipment.

FDA expects firms to have written general procedures on how cleaning processes will be validated.

FDA expects the general validation procedures to address who is responsible for performing and approving the validation study, the acceptance criteria, and when revalidation will be required.

FDA expects firms to prepare specific written validation protocols in advance for the studies to be performed on each manufacturing system or piece of equipment which should address such issues as sampling procedures, and analytical methods to be used including the sensitivity of those methods.

FDA expects firms to conduct the validation studies in accordance with the protocols and to document the results of studies.

FDA expects a final validation report which is approved by management and which states whether or not the cleaning process is valid. The data should support a conclusion that residues have been reduced to an "acceptable level."

This section is not specific to the maintenance department. However, knowing what needs to be cleaned from the equipment, (e.g., maintenance supplies or residue) may determine what type of cleaning the equipment must receive. This cleaning is different from the cleaning to remove bacteria or other contaminants.

Animal and Pest Control

This next section specifically discusses pest control, yet it actually is discussing animal control, since all animals are prohibited from any area of the plant where product contamination could possibly result.

Pest control.

No pests shall be allowed in any area of a food plant. Guard or guide dogs may be allowed in some areas of a plant if the presence of the dogs is unlikely to result in contamination of food, food-contact surfaces, or food-packaging materials. Effective measures shall be taken to exclude pests from the processing areas and to protect against the contamination of food on the premises by pests. The use of insecticides or rodenticides is permitted only under precautions and restrictions that will protect against the contamination of food, food- contact surfaces, and food-packaging materials.

General Sanitation

Additional FDA requirements related to the process and equipment include the following:

Equipment and utensils and finished food containers shall be maintained in an acceptable condition through appropriate cleaning and sanitizing, as necessary. Insofar as necessary, equipment shall be taken apart for thorough cleaning.

All food manufacturing, including packaging and storage, shall be conducted under such conditions and controls as are necessary to minimize the potential for the growth of microorganisms, or for the contamination of food. One way to comply with this requirement is careful monitoring of physical factors to ensure that mechanical breakdowns, time delays, temperature fluctuations, and other factors do not contribute to the decomposition or contamination of food.

By itself, this regulation favors a good PM program to eliminate unplanned downtime. As mentioned, if the equipment is unreliable, then alternative approaches to insure the quality of the product must be determined.

Work-in-process shall be handled in a manner that protects against contamination.

Effective measures shall be taken to protect finished food from contamination by raw materials, other ingredients, or refuse.

> When raw materials, other ingredients, or refuse are unprotected, they shall not be handled simultaneously in a receiving, loading, or shipping area if that handling could result in contaminated food. Food transported by conveyor shall be protected against contamination as necessary.
>
> Equipment, containers, and utensils used to convey, hold, or store raw materials, work-in-process, rework, or food shall be constructed, handled, and maintained during manufacturing or storage in a manner that protects against contamination.
>
> Effective measures shall be taken to protect against the inclusion of metal or other extraneous material in food. Compliance with this requirement may be accomplished by using sieves, traps, magnets, electronic metal detectors, or other suitable effective means.

This regulation also favors a strong PM program; all wear particles or points of wear should be identified and inspected on a regular basis to insure no contamination of the product is taking place.

> Mechanical manufacturing steps such as washing, peeling, trimming, cutting, sorting and inspecting, mashing, dewatering, cooling, shredding, extruding, drying, whipping, defatting, and forming shall be performed so as to protect food against contamination. Compliance with this requirement may be accomplished by providing adequate physical protection of food from contaminants that may drip, drain, or be drawn into the food. Protection may be provided by adequate cleaning and sanitizing of all food-contact surfaces, and by using time and temperature controls at and between each manufacturing step.
>
> Batters, breading, sauces, gravies, dressings, and other similar preparations shall be treated or maintained in such a manner that they are protected against contamination. Compliance with this requirement may be accomplished by any effective means, including one or more of the following:
>
> > Providing adequate physical protection of components from contaminants that may drip, drain, or be drawn into them.

Both of the previous regulations place importance on not over-servicing a component to the point that excess oil or grease might contaminate the product. Proper PM procedures should be specified and followed.

Calibration

> The calibration of instruments, apparatus, gauges, and recording devices at suitable intervals in accordance with an established written program containing specific directions, schedules, limits for accuracy and precision, and provisions for remedial action in the event accuracy and/or precision limits are not met. Instruments, apparatus, gauges, and recording devices not meeting established specifications shall not be used.

This regulation is in the PM section, since most organizations view the calibration control as part of the PM program. However, the FDA requirements are clear on the importance of this inspection. If the device has to be removed from service, the result will be an equipment shutdown or excessive stocking of spare parts – both of which are expensive.

Metal Inclusion

In a further clarification of the inspection process, the FDA has additional guidelines where food is not going through a metal detection device, but still has the possibility of contamination. The guidelines say:

> If the product will not be run through such a device, you should have procedures to periodically check the processing equipment for damage or lost parts at each processing step where "metal inclusion" is identified as a significant hazard. In this case you should identify those processing steps as CCPs. It would not ordinarily be necessary to identify these steps as CCPs in addition to identifying a final metal detection or separation step as a CCP.
>
> Visually inspecting equipment for damage or missing parts may only be feasible with relatively simple equipment, such as band saws, small orbital blenders, and wire-mesh belts. Other, more complex, equipment may contain to many parts, some of which

may not be readily visible, to make such visual inspection reliable in a reasonable time period.

Good PM inspections at a frequent interval can help in achieving compliance with this standard.

Preventive Maintenance

In another section of the regulations, the FDA dealt with Juice Concentrates manufacturing and highlighted this requirement:

> Preventive maintenance programs should be established to ensure the proper functioning of the equipment and integrity of the food contact surfaces. Periodic inspection of containers and equipment and periodic replacement of gaskets and flexible hoses are examples of preventive maintenance programs.

While the statement is brief, it is quite clear that the FDA expects to find an effective PM program in these plants.

Drugs and Drug-Related Products

In further FDA guidelines for the manufacture of drugs or drug related products, the regulations state:

This system includes the measures and activities which provide an appropriate physical environment and resources used in the production of the drugs or drug products. It includes:

a) Buildings and facilities along with maintenance;

b) Equipment qualifications (installation and operation); equipment calibration and preventative maintenance; and cleaning and validation of cleaning processes as appropriate. Process performance qualification will be evaluated as part of the inspection of the overall process validation which is done within the system where the process is employed; and,

c) Utilities that are not intended to be incorporated into the product such as HVAC, compressed gases, steam and water systems.

These regulations specify the PM program, the calibration program, and proper design and installation of the equipment. However, it goes further in specifying the utility systems, which help insure the uptime of the lines and the quality of the product.

Drug manufacturers' test labs are also included in the regulations.

Drug Laboratory Control System

For each of the following, the firm should have written and approved procedures and documentation resulting there from. The firm's adherence to written procedures should be verified through observation whenever possible. These areas are not limited only to finished products, but may also incorporate components and in-process materials. These areas may indicate deficiencies not only in this system but also in other systems that would warrant expansion of coverage. When this system is selected for coverage in addition to the Quality System, all areas listed below should be covered; however, the depth of coverage may vary depending upon inspectional findings.

- training/qualification of personnel
- adequacy of staffing for laboratory operations
- adequacy of equipment and facility for intended use
- calibration and maintenance programs for analytical instruments and equipment
- validation and security of computerized or automated processes
- reference standards; source, purity and assay, and tests to establish equivalency to current official reference standards as appropriate
- system suitability checks on chromatographic systems (e.g., GC or HPLC)
- specifications, standards, and representative sampling plans
- adherence to the written methods of analysis
- validation/verification of analytical methods
- control system for implementing changes in laboratory operations
- required testing is performed on the correct samples
- documented investigation into any unexpected discrepancy
- complete analytical records from all tests andsummaries of results
- quality and retention of raw data (e.g., chromatograms and spectra)
- correlation of result summaries to raw data; presence of unused data
- adherence to an adequate Out of Specification (OOS) procedure which includes timely completion of the investigation

- adequate reserve samples; documentation of reserve sample examination
- stability testing program, including demonstration of stabil ity indicating capability of the test methods

Guidelines for Cosmetic Manufacturers

While considering the regulations for the drug industry, it might be beneficial to examine the guidelines for the cosmetic industry.

1. Building and Facilities. Check whether:

Buildings used in the manufacture or storage of cosmetics are of suitable size, design and construction to permit unobstructed placement of equipment, orderly storage of materials, sanitary operation, and proper cleaning and maintenance.

Floors, walls and ceilings are constructed of smooth, easily cleanable surfaces and are kept clean and in good repair.

Fixtures, ducts and pipes are installed in such a manner that drip or condensate does not contaminate cosmetic materials, utensils, cosmetic contact surfaces of equipment, or finished products in bulk.

Lighting and ventilation are sufficient for the intended operation and comfort of personnel.

Water supply, washing and toilet facilities, floor drainage and sewage system are adequate for sanitary operation and cleaning of facilities, equipment and utensils, as well as to satisfy employee needs and facilitate personal cleanliness.

2. Equipment. Check whether:

Equipment and utensils used in processing, holding, transferring and filling are of appropriate design, material and workmanship to prevent corrosion, buildup of material, or adulteration with lubricants, dirt or sanitizing agent.

Utensils, transfer piping and cosmetic contact surfaces of equipment are well-maintained and clean and are sanitized at appropriate intervals.

Cleaned and sanitized portable equipment and utensils are stored and located, and cosmetic contact surfaces of equipment are covered, in a manner that protects them from splash, dust or other contamination.

3. Personnel. Check whether:

The personnel supervising or performing the manufacture or control of cosmetics has the education, training and/or experience to perform the assigned functions.

Persons coming into direct contact with cosmetic materials, finished products in bulk or cosmetic contact surfaces, to the extent necessary to prevent adulteration of cosmetic products, wear appropriate outer garments, gloves, hair restraints etc., and maintain adequate personal cleanliness.

Consumption of food or drink, or use of tobacco is restricted to appropriately designated areas.

4. Facilities and Equipment System.

This system includes the measures and activities which provide an appropriate physical environment and resources used in the production of the drugs or drug products. It includes:

a) Buildings and facilities along with maintenance;

b) Equipment qualifications (installation and operation); equipment calibration and preventative maintenance; and cleaning and validation of cleaning processes as appropriate. Process performance qualification will be evaluated as part of the inspection of the overall process validation which is done within the system where the process is employed; and,

c) Utilities that are not intended to be incorporated into the product such as HVAC, compressed gases, steam and water systems.

Additional FDA Guidelines

For each of the following, the firm should have written and approved procedures and documentation resulting there from. The firm's adherence to written procedures should be verified through observation whenever possible. These areas may indicate deficiencies not only in this system but also in other systems that would warrant expansion of coverage. When this system is selected for coverage in addition to the Quality System, all areas listed below should be covered; however, the depth of coverage may vary depending upon inspectional findings.

1. Facilities

- cleaning and maintenance
- facility layout and air handling systems for prevention of cross-contamination (e.g.penicillin, beta-lactams, steroids, hormones cytotoxics, etc.)
- specifically designed areas for the manufacturing operations performed by the firm to prevent contamination or mix-ups
- general air handling systems
- control system for implementing changes in the building
- lighting, potable water, washing and toilet facilities, sewage refuse disposal
- sanitation of the building, use of rodenticides, fungicides, insecticides, cleaning and sanitizing agents

2. Equipment

- equipment installation and operational qualification where appropriate
- adequacy of equipment design, size, and location
- equipment surfaces should not be reactive, additive, or absorptive
- appropriate use of equipment operations substances, (lubricoolants, refrigerants, etc.) contacting products/containers/etc.
- cleaning procedures and cleaning validation
- controls to prevent contamination, particularly with any pesticides or any other toxic materials, or other drug or non-drug chemicals

- — qualification, calibration and maintenance of storage equip ment, such as refrigerators and freezers for ensuring that standards, raw materials, reagents, etc. are stored at the proper temperatures
- — equipment qualification, calibration and maintenance, including computer qualification/validation and security
- — control system for implementing changes in the equipment
- — equipment identification practices (where appropriate)
- — documented investigation into any unexpected discrepancy

While these requirements are strikingly similar to the drug industry (they are related), they again highlight the need for an effective preventive maintenance and record keeping program.

Medicated Feeds

The FDA standards also extend beyond processes designed for human consumption. The following is excerpted from a section on medicated feeds.

CURRENT GOOD MANUFACTURING PRACTICE FOR MEDICATED FEEDS

Subpart F — Facilities and Equipment

- *225.120 Buildings and grounds.*

Buildings used for production of medicated feed shall provide adequate space for equipment, processing, and orderly receipt and storage of medicated feed. Areas shall include access for routine maintenance and cleaning of equipment. Buildings and grounds shall be constructed and maintained in a manner to minimize vermin and pest infestation.

Comment: Buildings and grounds should be suitable for their use and facilitate the production of feeds that are proper in all respects. Construction, maintenance, and upkeep should provide protection from the elements and pests. The key elements are suitability and good housekeeping.

- *225.130 Equipment.*

Equipment shall be capable of producing a medicated feed of intended potency and purity, and shall be maintained in a reasonably clean and orderly manner. Scales and liquid metering devices shall be accurate and of suitable size, design, construction, precision, and accuracy for their intended purposes. All equipment shall be designed, constructed, installed, and maintained so as to facilitate inspection and use of cleanout procedure(s).

Comment: Equipment must be suitable for its purpose. It must be of proper size and design for the function performed to have the inherent needed capability to produce good products. Good maintenance is equally important. The key elements are accuracy, capability, and good maintenance.

Self-inspections of non-registered feed manufacturers should cover at least the following areas:

A. Facilities and Equipment

- *225.120 Buildings and grounds:* Is there adequate space for equipment, and processing and storage of medicated feeds? Does construction and maintenance minimize vermin and pest infestation?

- *225.130 Equipment:* Is equipment capable of producing a medicated feed of intended potency and integrity? Are adequate cleanout procedures used to avoid unsafe contamination of medicated and non-medicated feeds? Such procedures may include physical cleanout, flushing, sequencing of production, and similar actions. Are scales and metering devices accurate and suitable for their intended purposes?

- *225.135 Work and Storage Areas:* Are work areas, drug storage, and equipment free of pesticides, fertilizers and other toxic substances that could contaminate feeds?

Laboratory Control Systems

Another perspective on PM programs can be derived from the FDA's Laboratory Control System.

For each of the following, the firm should have written and approved procedures and documentation resulting there from. The firm's adherence to written procedures should be verified through observation whenever possible. These areas are not limited only to finished products, but may also incorporate components and in-process materials. These areas may indicate deficiencies not only in this system but also in other systems that would warrant expansion of coverage. When this system is selected for coverage in addition to the Quality System, all areas listed below should be covered; however, the depth of coverage may vary depending upon inspectional findings.

- training/qualification of personnel
- adequacy of staffing for laboratory operations
- adequacy of equipment and facility for intended use
- calibration and maintenance programs for analytical instruments and equipment
- validation and security of computerized or automated processes
- adherence to an adequate Out of Specification (OOS) procedure which includes timely completion of the investigation
- adequate reserve samples; documentation of reserve sample examination
- stability testing program, including demonstration of stability indicating capability of the test methods

In the laboratory example, different issues were highlighted, including the training and staffing needed to accomplish the required tasks. While laboratory equipment is more technical in nature than some plant equipment, the maintenance requirements still exist. So a high level of skill may be required to properly calibrate and service the equipment.

Medical Regulations

Before leaving the drug and cosmetic theme. let us consider some of the medical regulations. These address issues facing hospitals and inspection of their equipment. The first is the FDA's guidelines on X-ray machines.

X-ray Machines.

Responsibility for Defects and Noncompliances. The Diagnostic X-ray Performance Standard specifies certain limits of responsibility for manufacturers and assemblers of x-ray equipment. Assemblers are responsible only for noncompliances that are "attributed solely to improper assembly or installation..." caused by improperly following the instructions provided by the manufacturer. Manufacturers are responsible for noncompliance's caused by improper assembly only if adequate instructions were not provided to the assembler (1020.30(c)). The performance standard does not specifically address the limits of responsibility regarding equipment age or user responsibility.

Manufacturers are required by the performance standard to provide purchasers with a schedule of maintenance necessary to keep the x-ray equipment in compliance with the performance standard. The regulations require manufacturers to provide a maintenance schedule because it is unreasonable to expect x-ray equipment to meet certain performance requirements if proper maintenance is not performed. After the first maintenance is performed or after the time it should have been performed, the assembler may no longer be responsible for requirements affected by proper adherence to the maintenance schedule. Some assemblers of older certified equipment will correct the noncompliance and bill the owner rather than attempting to refute responsibility for the noncompliance. This practice frequently upsets the x-ray system owner since he believes this work should have been performed free of charge.

Evidence that would exempt manufacturers/assemblers from responsibility includes:

1. Failure by the user to follow the manufacturer's prescribed maintenance schedule for those items requiring periodic adjustment.

2. Photographs or other documentation (written description) of physical damage to the x-ray system which was due to abuse.

The manufacturer/assembler may be held responsible if the user has failed to follow the maintenance schedule but the facility has documented continued compliance problems with the system beginning in the warranty period.

Items that may require periodic adjustment under a manufacturer's maintenance schedule include:

a) linearity
b) x-ray field/light field alignment
c) PBL sizing
d) illuminance
e) entrance exposure rate

f) fluoroscopic alignment
g) spot film alignment
h) indication of technique factors
i) signal and warning lights

Some items require adjustment on a time basis while others require adjustment at time of a tube reloading or bulb change in the collimator lamp. The individual maintenance schedule must be checked to determine the applicable situation and time interval.

While it is true that most hospitals outside contract the maintenance on this equipment, it is still in their best interests to understand the regulations and insure their contractors are in compliance.

The PM programs for hospitals can get quite specific. The following is just one example.

> FDA recommends the following actions to prevent deaths and injuries from entrapment in hospital bed side rails:

Inspect all hospital bed frames, bed side rails, and mattresses as part of a regular maintenance program to identify areas of possible entrapment. Regardless of mattress width, length, and/or depth, alignment of the bed frame, bed side rail, and mattress should leave no gap wide enough to entrap a patient's head or body. Be aware that gaps can be created by movement or compression of the mattress which may be caused by patient weight, patient movement, or bed position.

While this regulation may seem at first to be trivial, consider the effort needed for a hospital staff to comply with it.

EPA Regulations

The EPA has its own set of regulations that are important to maintenance departments.

Indoor Air Quality

An important set of EPA regulations govern indoor air quality. For example, the following regulations provide a checklist for schools.

Maintenance Supplies

Maintenance supplies may emit air contaminants during use and storage. Products low in emissions are preferable. However, a product that is low in emissions is not necessarily better if it is more hazardous, despite the lower emissions, if it has to be used more often or at a higher strength. Examples of maintenance supplies that may contribute to indoor air quality (IAQ) problems include:

- Caulks
- Solvents
- Paints
- Adhesives
- Sealants
- Cleaning Agents

Learn about your maintenance supplies.

- Review and become familiar with your maintenance supplies.
- Read labels and identify precautions regarding effects on indoor air or ventilation rate and requirements.
- If you make purchase decisions, or recommend products for purchase, confirm that supplies are safe to use.
- Ask vendors and manufacturers to help select the safest products available that can accomplish the job effectively.
- Follow good safty, handling, disposal, and storage practices.
- Develop appropriate procedures and have supplies available for spill control.
- Exhaust air from chemical and trash storage areas to the outdoors.
- Store chemical products and supplies in sealable, clearly labeled containers.
- Follow manufacturers' instructions for use of maintenance supplies.
- Follow manufacturers' instructions for disposal of chemicals, chemical-containing wastes, and containers.
- Establish maintenance practices that minimize occupant exposure to hazardous materials
- Substitute less- or non-hazardous materials where possible.
- Schedule work involving odorous or hazardous chemicals for periods when the school is unoccupied.
- Ventilate during and after use of odorous or hazardous chemicals.

Mold and Mildew

Many people have allergic reactions to mold and mildew. Mold and mildew can grow almost anywhere that offers a food source and a small amount of moisture, whether from leaks and spills or condensation. Mold and mildew do not require standing water in order to grow. The higher the relative humidity, the higher the probability of fungal growth.

Assemble the following tools before starting the activities:

- a small floor plan for taking notes.
- an instrument to measure relative humidity (e.g., sling psychrometer)

Inspect the building for signs of moisture, leaks, or spills.

- Check for moldy odors.
- Look for stains or discoloration on the ceiling, walls, or floor.
- Check cold surfaces (e.g., locations under windows and in corners formed by exterior walls, uninsulated cold water piping)
- Check areas where moisture is generated (e.g., locker rooms, bathrooms).
- Look for signs of water damage in:
- indoor areas in the vicinity of known roof or wall leaks;
- walls around leaky or broken windows;
- floors and ceilings under plumbing; and, duct interiors near humidifiers, cooling coils, and outdoor air intakes.

If you discover active leaks during your inspection, note their location(s) on your floor plan and repair them as quickly as possible.

Respond promptly when you see signs of moisture, or when leaks or spills occur.

- Clean and dry damp or wet building materials and furnishings
- Work with manufacturers of furnishings and building materials to learn recommended cleaning procedures and/or identify competent contractors who can clean damp materials.

Porous, absorbent building materials or furnishings, such as ceiling tiles, wall boards, floor coverings, etc., must be thoroughly dried and cleaned as soon as possible. In some cases these materials might have to be disinfected. If these materials can't be dried and cleaned within 24 hours, they may have to be replaced after the cause of the moisture problem has been corrected.

Prevent moisture condensation.

There are several methods to prevent condensation:

- Reduce the potential for condensation on cold surfaces (piping, exterior walls, roof, or floor) by adding insulation.

(Note: When installing insulation that has a vapor barrier, put the vapor barrier on the warm side of the insulation.).

- Raise the temperature of the air.
- Improve air circulation in the problem location.
- Decrease the amount of water vapor in the air.
- In dryer climates or winter, supply more outdoor ventilation air
- In humid climates or during humid times of the year, use a dehumidifier or desiccants to dry the air (for more information, obtain Appendix H from the IAQ Coordinator)
- Increase the capacity or operating schedule of existing exhaust fan(s); or add a local exhaust fan near the source of the water vapor

While these guidelines were written specifically for schools, they can apply to virtually any facility. It is beneficial for all facilities to consider these points while designing the PM program.

Road Maintenance

For a dramatically different look at a preventive maintenance issue, consider the EPA standards for road maintenance. While this doesn't affect all plants, many plants have internal roads and they are responsible for maintaining them.

Importance to Maintenance & Water Quality

Disturbances to unpaved roadway surfaces and ditches, and poor road surface drainage always result in deterioration of the road surface. This deterioration is the erosion which accounts for a large percentage of unpaved road maintenance costs and stream sedimentation.

Frequent, excessive, and unnecessary disturbances to the roadways are all too common because of political pressure from the public to continually blade roads, and the common practice of wholesale blading adopted by administrators and operators over the years. Proper and timely surface maintenance, selectively performed, will help reduce the amount of roadway being disturbed, and will reduce the amount and frequency of disturbance to the section of roadway requiring maintenance.

Proper, timely, and selective surface maintenance, which includes water disposal, prevents and minimizes erosion problems, thereby lengthening the life of the road surface which in turn lessens frequency and cost of maintenance. This will also decrease the amount of sediment carried into surface waters. Frequent and excessive disturbance of the roadway surface and ditches, and failure to direct surface water from the road surface to a drainage channel results in deterioration of the road surface, which leads to other roadway problems which may impair traffic flow and traffic safety, among other things.

As can be seen, improper road maintenance can have an impact on run-off standards and increase pollution, which impacts the pollution program for a plant. A PM program for a plant would, by regulation, include road maintenance.

Small Business Specialty Concerns

The following EPA regulations are specific to the farming industry. Although the guidelines are specific, they would be good to consider if such issues arise in plants. These guidelines are presented from the perspective of the EPA.

Vehicle and Machine Maintenance and Repair

Day-to-day maintenance and repair activities keep farm machinery and vehicles safe and reliable. Maintenance activities include oil and filter changes, battery replacement, and repairs including light metal machining.

Fluids

Potential wastes generated as a result of farm machinery and vehicle maintenance and repair activities can include used oil, spent fluids, spent batteries, asbestos brake pads and linings, metal machining wastes, spent organic solvents, and tires. These wastes have the potential to be released to the environment if not handled properly, stored in secure areas with secondary containment, and/or protected from exposure to weather. If released to the environment, the impact of these releases can be contamination of surface waters, groundwater, and soils, as well as toxic releases to the air.

Batteries

Farm operators have three options for managing used batteries: recycling through a supplier, recycling directly through a battery reclamation facility, or direct disposal. Most suppliers now accept spent batteries at the time of new battery purchase. While some waste batteries must be handled as hazardous waste, lead acid batteries are not considered hazardous waste as long as they are recycled. In general, recycling batteries may reduce the amount of hazardous waste stored at a farm, and thus reduce the farm's responsibilities under RCRA.

The following best management practices are recommended to prevent used batteries from impacting the environment prior to disposal:

- Place on pallets and label by battery type.
- Protect them from the weather with a tarp, roof, or other means.
- Store them on an open rack or in a watertight secondary containment unit to prevent leaks.
- Inspect batteries for cracks and leaks as they come to the farm.
- Neutralize acid spills and dispose of the resulting waste as hazardous if it sill exhibits a characteristic of a hazardous waste.
- Avoid skin contact with leaking or damaged batteries.

Machine Shop Wastes

The major hazardous wastes from metal machining are waste cutting oils, spent machine coolant, and degreasing solvents. Scrap metal can also be a component of hazardous waste produced at a machine shop. Material substitution and recycling are the two best means to reduce the volume of these wastes.

The preferred method of reducing the amount of waste cutting oils and degreasing solvents is to substitute with water-soluble cutting oils. If non-water-soluble oils must be used, recycling waste cutting oil reduces the potential environmental impact. Machine coolant can be recycled, either by an outside recycler, or through a number of in-house systems. Coolant recycling is most easily implemented when a standardized type of coolant is used throughout the shop. Reuse and recycling of solvents also is easily achieved, although it is generally done by a permitted recycler.

Most shops collect scrap metals from machining operations and sell these to metal recyclers. Metal chips that have been removed from the coolant by filtration can be included in the scrap metal collection. Wastes should be carefully segregated to facilitate reuse and recycling.

Water and Wastewater Preventive Maintenance Guidelines

This section from the EPA guidelines is of interest, since it recommends not just a preventive maintenance system, but also provides guidelines for setting up and administering the PM program. Consider the following as it applies to the water – wastewater organizations.

Description

Preventive maintenance involves the regular inspection, testing, and replacement or repair of equipment and operational systems. As a storm water best management practice (BMP), preventive maintenance should be used to monitor systems built to control storm water. These systems should be inspected to uncover cracks, leaks, and other conditions that could cause breakdowns or failures of storm water mitigation structures and equipment, which, in turn, could result in discharges of chemicals to surface waters either by direct overland flow or through storm drainage systems. A preventive maintenance program can prevent breakdowns and failures through adjustment, repair, or replacement of equipment before a major breakdown or failure occurs. Typically, a preventive maintenance program should include inspections of catch basins, storm water detention areas, and water quality treatment systems. With-out adequate maintenance, sediment and debris can quickly clog storm drainage facilities and render them useless.

Applicability

Preventive maintenance procedures and activities are applicable to almost all industrial facilities. This concept should be a part of a general good housekeeping program designed to maintain a clean and orderly work environment. Often the most effective first step towards preventing storm water pollution from industrial sites is to improve the facility's preventive maintenance and general good housekeeping methods. For many facilities, preventive maintenance to protect water quality is simply an extension of current plant preventive maintenance programs. Most plants already have preventive maintenance programs that provide some degree of environmental protection. Such programs could be expanded to include storm water considerations.

Advantages and Disadvantages

Preventive maintenance takes a proactive approach to storm water management and seeks to prevent problems before they occur. A preventive maintenance program can improve water quality by controlling pollutant discharges to surface water that would result from spills and leaks. Preventive maintenance programs can also save a facility money by reducing the likelihood of having a system breakdown and also by reducing the likelihood of funding costly cleanup projects. In addition, a preventive maintenance program can be an effective community relations tool. The primary limitations of implementing a preventive maintenance program include:

Cost.
Availability of trained preventive maintenance staff technicians. Management direction and staff motivation in expanding the preventive maintenance program to include storm water considerations.

Key Program Components

Elements of a good preventive maintenance program should include the following:

i. Identification of equipment or systems that may malfunction and cause spills or leaks, or may otherwise contaminate storm water runoff. Typical equipment to be inspect ed includes pipes, pumps, storage tanks and bins, pressure vessels, pressure release valves, process and material handling equipment, and storm water management devices.

ii. Establishment of schedules and procedures for routine inspections.
iii. Periodic testing of plant equipment for structural soundness.
iv. Prompt repair or replacement of defective equipment found during inspection and testing.
v. Maintenance of a supply of spare parts for equipment that needs frequent repairs.
vi. Use of an organized record-keeping system to schedule tests and document inspections.
vii. Commitment to ensure that records are complete and detailed, and that they record test results and follow-up actions.

Preventive maintenance inspection records should be kept with other visual inspection records.

Implementation

The key to properly implementing and tracking a preventive maintenance program is through the continual updating of maintenance records. Update records immediately after performing preventive maintenance or repairing an item and review them annually to evaluate the overall effectiveness of the program. Then refine the preventive maintenance procedures as necessary.

No quantitative data on the effectiveness of preventive maintenance as a BMP is available. However, it is intuitively clear that an effective preventive maintenance program will result in improved storm water discharge quality.

Costs

The major cost of implementing a preventive maintenance program on storm water quality is the staff time required to administer the program. Typically, this is a small incremental increase if a preventive maintenance program already exists at the facility.

While this section does discount the cost, the focus is the impact of a good PM program on compliance issues. Since the recommendations in the document are for a standard PM program, all water and wastewater organizations should have a PM program they are going to use in order to comply with the regulations.

Combined Sewer Systems

The following is an EPA guideline for combined sewer systems. Preventive maintenance clearly plays a key role in the effective operation of these systems. Note that a PM program is one of the minimum requirements for the legal operation of a combined sewer system.

Proper Operation and Maintenance

Description

Combined sewer systems (CSSs) are single-pipe sewer systems that convey sanitary wastewaters (domestic, commercial and industrial) and storm water runoff to a publicly owned treatment works. During periods of heavy rainfall, however, the sanitary wastewaters and storm waters can overflow the conveyance system and discharge directly to surface water bodies. This is called a combined sewer overflow (CSO). CSOs may contain high levels of suspended solids, biochemical oxygen demand (BOD), oil and grease, floatables, toxic pollutants, pathogenic microorganisms and other pollutants. These pollutants can exceed water quality standards and pose risks to human health, threaten aquatic species, and damage the waterways.

Because of the pollution potential from CSOs, EPA issued the CSO Control Policy on April 19, 1994. This policy states that permittees with CSSs that have CSOs should be able to provide, at a minimum, primary treatment and disinfection, when necessary, to 85 percent of the volume captured in a CSS on an annual average basis. The policy also includes nine minimum control requirements for inclusion in the CSO discharge permit. One of these minimum controls is proper operation and regular maintenance (O&M) programs for the sewer systems with CSOs.

Key Program Components

Proper O&M of combined sanitary sewers and overflows is not significantly different from that of sanitary sewer systems, with the objective being to maintain maximum flow to the wastewater treatment plant and to maximize either in-line storage capacity or detention upstream of the system inlets. There are several key components of an O&M program that a municipality/authority must provide to ensure proper O&M and to meet the minimum control requirement. These program components include:

- Scheduling routine inspections, maintenance and cleaning of the CSS, regulators and outfalls.
- Developing O&M reporting and record keeping systems with maintenance procedures and inspection reports.
- Providing training for O&M personnel.
- Reviewing the O&M program periodically to up-date and revise procedures as necessary.

These components are further described below.

Operational Review

Prior to developing an O&M program, the municipality should undertake an operational review of its system to inventory and assess existing facilities, operating conditions and maintenance practices. The municipality should have a complete plan of the collection system, showing all sewers and points where CSOs and outfalls are located. This plan should reference streets and other utilities to enable the maintenance crews to locate the structures and CSOs quickly. This plan may also aid in scheduling and planning the inspection and maintenance of the CSS system and overflows; for example, the regions or areas that are prone to flooding or premature overflows should be inspected first after a major storm.

It is worth noting that this paragraph requires an analysis of the existing O&M practices. This is important since many organizations really don't understand their maintenance work practices.

The nine minimum CSO control requirements include conducting a characterization of the CSS. This characterization should include documentation of overflow occurrences and correlation of these events with rainfall patterns (e.g., volume, intensity, duration). The results of the CSS characterization are critical to designing an O&M program that is effective in optimizing system operations. As part of these studies, it is important to measure actual system flows and the response to various operating and wet weather conditions.

This information will be critical during the development of specific operation and maintenance procedures that will be part of the O&M program. Municipalities may eventually be able to use data from their Long-term CSO Control Plans to supplement their O&M programs. As part of these plans, a system may conduct modeling of the integrated system (sewers, regulators, and treatment plant) to analyze operational improvements. These modeling efforts typically identify operational modifications that maximize storage and transport, provide improved treatment in the existing system, and decrease untreated CSO discharges. Because many municipalities will implement their O&M programs before their Long-term CSO Control Plans are completed, the results of the CSS modeling may not be available during the early phase of the O&M program. However, the O&M program should be updated periodically to address this type of additional information.

Record Keeping System

The O&M program should include a record keeping component. The record keeping system should document maintenance procedures through inspection reports. These reports should include information about when the system was inspected, and, if applicable, what maintenance action was taken, including the equipment used and the personnel involved. Geographical information systems (GIS) and desktop mapping may be useful in storing O&M data on the CSO system, as well as in developing a database of problem areas.

This paragraph specifically requires a record keeping system for the maintenance program. It would be beneficial to consider the section on computerized maintenance management systems when deciding on a recordkeeping system.

System Operating Procedures

Each municipality should have written policies, procedures, or protocols for training O&M personnel and should conduct periodic reviews and revisions of the O&M program. Some municipalities have reported that alternating crews between O&M and other functions has proven beneficial because it reduces the tedium of the work by making it less routine, and it promotes the cross-training of employees. Other municipalities prefer devoting personnel strictly to O&M because it keeps the work assignments simple.

This paragraph in the standard offers the municipality the option of going to a team-based environment with cross-trained labor or keeping the lines drawn between operations and maintenance. It is very progressive.

Training

The O&M Program should have established training goals, procedures, and schedules. Training should provide the maintenance personnel with an understanding of the CSS operations and system characteristics. Hands-on training illustrates the specific O&M procedure to those directly responsible for performing these activities. In addition, the nature of the O&M work may require employees to work in confined spaces or to be exposed to dangerous gases. Providing proper safety training, in accordance with Occupational Safety and Health Adminis-tration (OSHA) standards, is imperative. Safety programs should be reviewed, and, if necessary, updated periodically. Tide gates that require underwater inspection should only be inspected by a certified diver.

Routine-Maintenance Activities

Proper operation of the CSO system begins with proper operation and maintenance of the individual components - the regulators, tide gates, pump stations, sewer lines, and catch basins; and implementation of an organized plan that provides regular, consistent, and response-oriented O&M. In addition, operators must develop plans for determining where CSOs occur, and for conducting system-specific repairs to prevent future CSOs.

This paragraph focuses on the basic maintenance requirements of the components of the system. It is desirable from a technical standpoint to design preventive maintenance programs at the component level and then cascade them upward to the system level.

Pump Station Maintenance

Pump stations should be maintained to operate at the design conditions. Wet wells should be routinely cleaned because grit and solids deposition in the wet well can damage or restrict the flow of wastewater into the pump. Inadequate or improper pump station operation can lead to reduced storage and hydraulic capacity during wet weather, and, if the pumping capacity is severely restricted, dry weather overflows can result. In general, inadequate pumping capacity is caused by:

- Mechanical, electrical, or instrumentation problems.
- Changes in the upstream drainage area that cause storm runoff to exceed the originaldesign basis.
- Changes in the discharge piping (e.g., tying-in or manifolding with another pressuresystem) that creates more headloss in the discharge system.

If conditions upstream of the pump station (such as development) increase the flow above the design values, steps should be taken to upgrade the station to meet the increased flowrate. Pump station upgrading may include such items as:

- Installing new pumps and motors.
- Changing out impellers.
- Upgrading/changing pump controls to maximize use of all pumps during wetweather.
- Modifying system piping to improve thesystem head curve.
- Installing additional force main piping forwet weather pumping.

Depending on the complexity of the system, changes to the downstream discharge conditions that may affect the system head curve may require extensive study and should be evaluated on a case-by-case basis.

The exerpt above from the EPA guidelines helps the municipality to realize the importance of good maintenance and operational practices. It is not just prudent for a municipality to implement this program, it is also cost effective, since the fines for regulatory violations far exceed the cost to properly maintain the system.

Grain Industry

The following are P.M. Guidelines from the Grain Industry.

Preventive maintenance.

The employer shall implement preventive maintenance procedures consisting of:

> Regularly scheduled inspections of at least the mechanical and safety control equipment associated with dryers, grain stream processing equipment, dust collection equipment including filter collectors, and bucket elevators;
>
> Lubrication and other appropriate maintenance in accordance with manufacturers' recommendations, or as determined necessary by prior operating records.

The employer shall promptly correct dust collection systems which are malfunctioning or which are operating below designed efficiency. Additionally, the employer shall promptly correct, or remove from service, overheated bearings and slipping or misaligned belts associated with inside bucket elevators.

A certification record shall be maintained of each inspection, performed in accordance with this paragraph (m), containing the date of the inspection, the name of the person who performed the inspection and the serial number, or other identifier, of the equipment specified in paragraph (m)(1)(i) of this section that was inspected.

The employer shall implement procedures for the use of tags and locks which will prevent the inadvertent application of energy or motion to equipment being repaired, serviced, or adjusted, which could result in employee injury. Such locks and tags shall be removed in accordance with established procedures only by the employee installing them or, if unavailable, by his or her supervisor.

Grain stream processing equipment.

> The employer shall equip grain stream processing equipment (such as hammer mills, grinders, and pulverizers) with an effective means of removing ferrous material from the incoming grain stream.

The grain regulations become even more specific, detailing some of the actual PM procedures that are required.

> The control of dust and the control of ignition sources are the most effective means for reducing explosion hazards. Preventive maintenance is related to ignition sources in the same manner as housekeeping is related to dust control and should be treated as a major function in a facility.
>
> Equipment such as critical bearings, belts, buckets, pulleys, and milling machinery are potential ignition sources, and periodic inspection and lubrication of such equipment through a scheduled preventive maintenance program is an effective method for keeping equipment functioning properly and safely. The use of vibration detection methods, heat sensitive tape or other heat detection methods that can be seen by the inspector or maintenance person will allow for a quick, accurate, and consistent evaluation of bearings and will help in the implementation of the program.
>
> The standard does not require a specific frequency for preventive maintenance. The employer is permitted flexibility in determining the appropriate interval for maintenance provided that the effectiveness of the maintenance program can be demonstrated. Scheduling of preventive maintenance should be based on manufacturer's recommendations for effective operation, as well as from the employer's previous experience with the equipment. However, the employer's schedule for preventive maintenance should be frequent enough to allow for both prompt identification and correction of any problems concerning the failure or malfunction of the mechanical and safety control equipment associated with bucket elevators, dryers, filter collectors and magnets.
>
> The pressure-drop monitoring device for a filter collector, and the condition of the lagging on the head pulley, are examples

of items that require regularly scheduled inspections. A system of identifying the date, the equipment inspected and the maintenance performed, if any, will assist employers in continually refining their preventive maintenance schedules and identifying equipment problem areas. Open work orders where repair work or replacement is to be done at a designated future date as scheduled, would be an indication of an effective preventive maintenance program.

It is imperative that the prearranged schedule of maintenance be adhered to regardless of other facility constraints. The employer should give priority to the maintenance or repair work associated with safety control equipment, such as that on dryers, magnets, alarm and shut-down systems on bucket elevators, bearings on bucket elevators, and the filter collectors in the dust control system.

Benefits of a strict preventive maintenance program can be a reduction of unplanned downtime, improved equipment performance, planned use of resources, more efficient operations, and, most importantly, safer operations

ISO-9000, ISO-14000, and PM

The regulations considered thus far clearly show the necessity of preventive maintenance programs. The ISO standards are no different. The ISO-9000 standards clearly call out calibration and maintenance requirements. It is a fundamental fact that one cannot operate a factory and achieve a proper quality management system without a preventive maintenance program. Without proper tracking of the maintenance and calibration of certain equipment and tools, no one can achieve the ISO-9000 standard. The highlights of the ISO-9000 standard include the following:

The calibration procedures must be documented

The equipment calibration must be traceable to an outside standard

The reference standard must be traceable to an outside standard, which is approved

All non-conforming equipment must be controlled

There are provisions for certain utility supplies, such as compressed air, water, electricity, and chemicals that must be under control and in a system of verification, if they are part of or impact the production process.

The same is true of the ISO-14000 standard. The process and control equipment needs careful maintenance. In addition, the calibration requirements for pollution control equipment is a critical requirement.

Without a preventive maintenance program, proper calibration of equipment and instruments, and a system of traceability, the ISO standards can not be achieved. Unfortunately many registrars miss this link. Their approach is strictly from the production perspective, leaving the maintenance factor out of the equation. This is true more of the United States registrars. The European-based registrars tend to give maintenance more consideration during the audit. Hopefully more consistency in the way organizations are audited will lead to more ISO emphasis on the maintenance part of the organization.

CHAPTER 2

Inventory and Purchasing

This chapter examines how regulatory requirements impact the maintenance stores and purchasing function. For our purposes, the tool room and tool storage functions are grouped under the stores function.

Inventory Control

According to the ISO-9000 standards, stores must be secure. They should not be open to access by unauthorized persons. This standard is typically applied to production stores where any small change in the product can have an impact on the quality of the product.

However, maintenance stores are as important to product quality. Any spare parts that are in substandard condition upon leaving the store can negatively impact product quality. For example, if the stores in a machining operation are uncontrolled, bearings could be opened, left out of their box and simply kept on the shelf. This condition allows for their deterioration. A technician could then take the bearing, installs it in a spindle, and, upon restart, get chatter in the machining operation. The fault is a defective bearing, but the problem originates in the maintenance stores, where management allowed an individual inside the stores, who then opened the bearings and left them on the shelf to deteriorate.

While the standards are open to different interpretations, registrars have made controlled and manned maintenance stores part of the criteria for ISO-9000 certification.

The ISO-14000 standards focus on the environmental impact. Do these standards afffect the stores and purchasing functions? Yes, because maintenance uses chemicals for cleaning. Furthermore, the stores and purchasing functions are responsible for the proper handling and storage of all maintenance-related supplies and equipment. These includes spare parts for equipment that may be part of an environmentally-sensitive process. Any defective spare part or substandard spare part can affect the process' compliance with the standards. Also, if spare parts are shown in stock in a computer system, but when a critical situation develops requiring that spare part, it is found not to be in stock, a serious situation can develop. An out-of-stock situation can occur frequently in plants with uncontrolled or open stores. Without dedicated stores personnel, record keeping is often not up the necessary level of accuracy.

Control of Tools and Equipment

In addition to inventory control, the stores department usually has the responsibility for the control of tools and equipment not assigned to an individual employee. This responsibility helps insure that any tools and equipment ordered meet safety specifications. When tools and equipment are used on a job, stores personnel have the responsibility to check them for any unsafe condition when returned. In some cases, the stores personnel are trained to maintain the tools and equipment, making the repairs needed to correct any unsafe condition. With these thoughts in mind, consider some of the guidelines for maintenance tools and equipment.

Portable Stepladders

This standard designates the three types of step ladders that the regulation recognizes.

> Stepladders longer than 20 feet shall not be supplied. Stepladders as hereinafter specified shall be of three types:
>
> Type I - Industrial stepladder, 3 to 20 feet for heavy duty, suchas utilities, contractors, and industrial use.
>
> Type II - Commercial stepladder, 3 to 12 feet for medium duty, such as painters, offices, and light industrial use.
>
> Type III - Household stepladder, 3 to 6 feet for light duty, such as light household use.

The section that impacts the maintenance organization states "Ladders shall be maintained in good condition at all times, the joint between the steps and side rails shall be tight, all hardware and fittings securely attached, and the movable parts shall operate freely without binding or undue play." By directly using the ladders, the maintenance department therefore become responsible for the inspection of the ladders.The maintenance department may choose to inspect the ladder each time it is used. The ladder may also be part of a weekly or monthly inspection, depending on its usage. Items mentioned specifically for inspection include the following:

Metal bearings of locks, wheels, pulleys, etc., shall be frequently lubricated.

Frayed or badly worn rope shall be replaced.

Safety feet and other auxiliary equipment

Ladders shall be inspected frequently and those which have developed defects shall be withdrawn from service for repair or destruction and tagged or marked as "Dangerous, Do Not Use."

Rungs should be kept free of grease and oil.

Metal Ladders.

Metal ladders have additional items for inspection and service. They include:

If a ladder is involved in any of the following, immediate inspection is necessary:

Ladders should be cleaned of oil, grease, or slippery materials. They should be cleaned with a solvent or steam cleaning. Ladders having defects are to be marked and taken out of service until repaired by either the maintenance department or the manufacturer.

If ladders tip over, inspect ladder for side rails dents or bends, or excessively dented rungs; check all rung-to-side-rail connections; check hardware connections; check rivets for shear.

All ladders shall be maintained in a safe condition. All ladders shall be inspected regularly, with the intervals between inspections being determined by use and exposure.

Additional Ladder Guidelines.

The regulations do not specify a specific inspection frequency. Instead,,they let the frequency depend on the usage of the ladder and the exposure to any type of environment that would accelerate the degradation of the ladder. In most cases, it is best to inspect the ladder before usage. Such an inspection is not time consuming.

The regulations specify that "good safe practices in the use and care of ladder equipment must be employed by the users.". The regulations then give a rule for proper setup of the ladder, stating, "A simple rule for setting up a ladder at the proper angle is to place the base a distance from the vertical wall equal to one-fourth the working length of the ladder."

In order to help an employer and employees stay in compliance, any work involving a ladder should contain inspection and setup reminders as a line in the work order plan. If the CMMS being utilized has a tool control module or a procedure library, these notes can be set up and then attached to each job.

Manlifts

Manlifts are power operated devices utilized to safely access repair areas that are elevated. They are common in most plants and facilities today. Since the use of manlifts is common, the regulations have specific guidelines for safe use. The guidelines include the following:

> All manlifts shall be inspected by a competent designated person at intervals of not more than 30 days. Limit switches shall be inspected and tested weekly. Manlifts found to be unsafe shall not be operated until properly repaired. The inspection of a manlift shall include, at a minimum, the following items:
>
> Steps.
>
> Step Fastenings.
>
> Rails.
>
> Rail Supports and Fastenings.
>
> Rollers and Slides.

Belt and Belt Tension.

Handholds and Fastenings.

Floor Landings.

Guardrails.

Lubrication.

Limit Switches.

Warning Signs and Lights.

Illumination.

Drive Pulley.

Bottom (boot) Pulley and Clearance.

Pulley Supports.

Motor.

Driving Mechanism.

Brake.

Electrical Switches.

Vibration and Misalignment.

"Skip" on up or down run when mounting step (indicating worn gears).

As with previous discussed equipment, the inspection record shall be kept for each inspection. The inspection data must include:

the date of the inspection,

the signature of the person who performed the inspection

the equipment number of the manlift which was inspected

The inspection records must be provided to an OSHA inspector upon request.

Slings

Slings of all types are used during the performance of a large portion of maintenance activities such as hoisting and transportation of major spares during repairs. This section looks at those regulatory requirements that are designed to insure slings are safe before they are used. The stores personnel may make the indicated inspections before the slings are issued to maintenance personnel. At a minimum, maintenance personnel should carry out their own inspection of the sling prior to and, sometimes, during its use.

The general regulations for slings require the following:

Slings that are damaged or defective shall not be used.

Slings shall not be shortened with knots or bolts or other makeshift devices.

Sling legs shall not be kinked.

Slings shall not be loaded in excess of their rated capacities.

Slings used in a basket hitch shall have the loads balanced to prevent slippage.

Slings shall be securely attached to their loads.

Slings shall be padded or protected from the sharp edges of their loads.

Suspended loads shall be kept clear of all obstructions.

All employees shall be kept clear of loads about to be lifted and of suspended loads.

Hands or fingers shall not be placed between the sling and its load while the sling is being tightened around the load.

Alloy Steel Chain Slings.

Chain slings have additional requirements. The regulations state:

Alloy steel chain slings shall have permanently affixed durable identification stating size, grade, rated capacity, and reach.

Attachments.

Hooks, rings, oblong links, pear shaped links, welded or mechanical coupling links or other attachments shall have a rated capacity at least equal to that of the alloy steel chain with which they are used or the sling shall not be used in excess of the rated capacity of the weakest component.

Makeshift links or fasteners formed from bolts or rods, or other such attachments, shall not be used.

Inspections.
In addition to the inspection points mentioned previously, a thorough periodic inspection of alloy steel chain slings in use shall be made on a regular basis, to be determined on the basis of (A) frequency of sling use; (B) severity of service conditions; (C) nature of lifts being made; and (D) experience gained on the service life of slings used in similar circumstances. Such inspections shall in no event be at intervals greater than once every 12 months.

The employer shall make and maintain a record of the most recent month in which each alloy steel chain sling was thoroughly inspected, and shall make such record available for examination.

The thorough inspection of alloy steel chain slings shall be performed by a competent person designated by the employer, and shall include a thorough inspection for wear, defective welds, deformation and increase in length. Where such defects or deterioration are present, the sling shall be immediately removed from service.

Proof testing.
The employer shall ensure that before use, each new, repaired, or reconditioned alloy steel chain sling, including all welded components in the sling assembly, shall be proof tested by the sling manufacturer or equivalent entity, in accordance with paragraph 5.2 of the American Society of Testing and Materials Specification A391-65, which is incorporated by reference as specified in Sec. 1910.6 (ANSI G61.1-1968). The employer shall retain a certificate of the proof test and shall make it available for examination.

Sling use.
Alloy steel chain slings shall not be used with loads in excess of their rated capacities.

Safe operating temperatures.
Alloy steel chain slings shall be permanently removed from service if they are heated above 1000 deg. F. When exposed

to service temperatures in excess of 600 deg. F, maximum working load limits shall be reduced in accordance with the chain or sling manufacturer's recommendations.

Repairing and reconditioning alloy steel chain slings.

Worn or damaged alloy steel chain slings or attachments shall not be used until repaired. When welding or heat testing is performed, slings shall not be used unless repaired, reconditioned and proof tested by the sling manufacturer or an equivalent entity.

Mechanical coupling links or low carbon steel repair links shall not be used to repair broken lengths of chain.

Deformed attachments.

Alloy steel chain slings with cracked or deformed master links, coupling links or other components shall be removed from service.

Slings shall be removed from service if hooks are cracked, have been opened more than 15 percent of the normal throat opening measured at the narrowest point or twisted more than 10 degrees from the plane of the unbent hook.

Wire Rope Slings.

Wire rope slings have some specific requirements. The following are some specifics from the standards.

Sling use.

Wire rope slings shall not be used with loads in excess of their rated capacities.

Minimum sling lengths.

Cable laid and 6 X 19 and 6 X 37 slings shall have a minimum clear length of wire rope 10 times the component rope diameter between splices, sleeves or end fittings.

Braided slings shall have a minimum clear length of wire rope 40 times the component rope diameter between the loops or end fittings.

Cable laid grommets, strand laid grommets and endless slings shall have a minimum circumferential length of 96 times their body diameter.

Safe operating temperatures.

Fiber core wire rope slings of all grades shall be permanently removed from service if they are exposed to temperatures in excess of 200 deg. F. When nonfiber core wire rope slings of any grade are used at temperatures above 400 deg. F or below minus 60 deg. F, recommendations of the sling manufacturer regarding use at that temperature shall be followed.

End attachments.

Welding of end attachments, except covers to thimbles, shall be performed prior to the assembly of the sling.

All welded end attachments shall not be used unless proof tested by the manufacturer or equivalent entity at twice their rated capacity prior to initial use. The employer shall retain a certificate of the proof test, and make it available for examination.

Removal from service.

Wire rope slings shall be immediately removed from service if any of the following conditions are present:

Ten randomly distributed broken wires in one rope lay, or five broken wires in one strand in one rope lay.

Wear or scraping of one-third the original diameter of outside individual wires.

Kinking, crushing, bird caging or any other damage resulting in distortion of the wire rope structure.

Evidence of heat damage.

End attachments that are cracked, deformed or worn.

Hooks that have been opened more than 15 percent of the normal throat opening measured at the narrowest point or twisted more than 10 degrees from the plane of the unbent hook.

Corrosion of the rope or end attachments.

Metal Mesh Slings

A third type of sling is the metal mesh slings. These have additional regulations. They are listed in the following section.

Sling marking.

Each metal mesh sling shall have permanently affixed to it a durable marking that states the rated capacity for vertical basket hitch and choker hitch loadings.

Handles.

Handles shall have a rated capacity at least equal to the metal fabric and exhibit no deformation after proof testing.

Attachments of handles to fabric.

The fabric and handles shall be joined so that:

The rated capacity of the sling is not reduced.
The load is evenly distributed across the width of the fabric.
Sharp edges will not damage the fabric.

Sling coatings.

Coatings which diminish the rated capacity of a sling shall not be applied.

Sling testing.

All new and repaired metal mesh slings, including handles, shall not be used unless proof tested by the manufacturer or equivalent entity at a minimum of 1 1/2 times their rated capacity. Elastomer impregnated slings shall be proof tested before coating.

Proper use of metal mesh slings.

Metal mesh slings shall not be used to lift loads in excess of their rated capacities as prescribed in Table N-184-15. Slings not included in this table shall be used only in accordance with the manufacturer's recommendations.

Safe operating temperatures.
Metal mesh slings which are not impregnated with elastomers may be used in a temperature range from minus 20 deg. F to plus 550 deg. F without decreasing the working load limit. Metal mesh slings impregnated with polyvinyl chloride or neoprene may be used only in a temperature range from zero degrees to plus 200 deg. F. For operations outside these temperature ranges or for metal mesh slings impregnated with other materials, the sling manufacturer's recommendations shall be followed.

Repairs.
Metal mesh slings which are repaired shall not be used unless repaired by a metal mesh sling manufacturer or an equivalent entity.
Once repaired, each sling shall be permanently marked or tagged, or a written record maintained, to indicate the date and nature of the repairs and the person or organization that performed the repairs. Records of repairs shall be made available for examination.

Removal from service.
Metal mesh slings shall be immediately removed from service if any of the following conditions are present:

A broken weld or broken brazed joint along the sling edge.

Reduction in wire diameter of 25 per cent due to abrasion or 15 per cent due to corrosion.

Lack of flexibility due to distortion of the fabric.

Distortion of the female handle so that the depth of the slot is increased more than 10 per cent.

Distortion of either handle so that the width of the eye is decreased more than 10 per cent.

A 15 percent reduction of the original cross sectional area of metal at any point around the handle eye.

Distortion of either handle out of its plane.

Natural and Synthetic Fiber Rope Slings

An additional type of lifting sling is the fiber rope. The following are the guidelines for safe use of natural or synthetic fiber rope slings.

Sling use.
Fiber rope slings made from conventional three strand construction fiber rope shall not be used with loads in excess of their rated capacities.

Safe operating temperatures.
Natural and synthetic fiber rope slings, except for wet frozen slings, may be used in a temperature range from minus 20 deg. F to plus 180 deg. F without decreasing the working load limit. For operations outside this temperature range and for wet frozen slings, the sling manufacturer's recommendations shall be followed.

Splicing.
Spliced fiber rope slings shall not be used unless they have been spliced in accordance with the following minimum requirements and in accordance with any additional recommendations of the manufacturer:

In manila rope, eye splices shall consist of at least three full tucks, and short splices shall consist of at least six full tucks, three on each side of the splice center line.

In synthetic fiber rope, eye splices shall consist of at least four full tucks, and short splices shall consist of at least eight full tucks, four on each side of the center line.

Strand end tails shall not be trimmed flush with the surface of the rope immediately adjacent to the full tucks. This applies to all types of fiber rope and both eye and short splices. For fiber rope under one inch in diameter, the tail shall proj ect at least six rope diameters beyond the last full tuck. For fiber rope one inch in diameter and larger, the tail shall project at least six inches beyond the last full tuck. Where a projecting tail interferes with the use of the sling, the tail shall be tapered and spliced into the body of the rope using at least two additional tucks (which will require a tail length of approximately six rope diameters beyond the last full tuck).

Fiber rope slings shall have a minimum clear length of rope between eye splices equal to 10 times the rope diameter.

Knots shall not be used in lieu of splices.

Clamps not designed specifically for fiber ropes shall not be used for splicing.

For all eye splices, the eye shall be of such size to provide an included angle of not greater than 60 degrees at the splice when the eye is placed over the load or support.

End attachments.
Fiber rope slings shall not be used if end attachments in contact with the rope have sharp edges or projections.

Removal from service.
Natural and synthetic fiber rope slings shall be immediately removed from service if any of the following conditions are present:

Abnormal wear.

Powdered fiber between strands.

Broken or cut fibers.

Variations in the size or roundness of strands.

Discoloration or rotting.

Distortion of hardware in the sling.

Repairs.
Only fiber rope slings made from new rope shall be used. Use of repaired or reconditioned fiber rope slings is prohibited.

nthetic Web Slings

Synthetic web slings are separated from the other synthetic slings, ıe to the web construction. The following are some specifics.

Sling identification.
Each sling shall be marked or coded to show the rated capacities for each type of hitch and type of synthetic web material.

Webbing.
Synthetic webbing shall be of uniform thickness and width and selvage edges shall not be split from the webbing's width.

Fittings.
Fittings shall be:

Of a minimum breaking strength equal to that of the sling; and

Free of all sharp edges that could in any way damage the webbing.

Attachment of end fittings to webbing and formation of eyes. Stitching shall be the only method used to attach end fittings to webbing and to form eyes. The thread shall be in an even pat tern and contain a sufficient number of stitches to develop the full breaking strength of the sling.

Environmental conditions.
When synthetic web slings are used, the following precautions shall be taken:

Nylon web slings shall not be used where fumes, vapors, sprays, mists or liquids of acids or phenolics are present.

Polyester and polypropylene web slings shall not be used where fumes, vapors, sprays, mists or liquids of caustics are present.

Web slings with aluminum fittings shall not be used where fumes, vapors, sprays, mists or liquids of caustics are present.

Safe operating temperatures. Synthetic web slings of polyester and nylon shall not be used at temperatures in excess of 180 deg. F. Polypropylene web slings shall not be used at temperatures in excess of 200 deg. F.

Repairs.
Synthetic web slings which are repaired shall not be used unless repaired by a sling manufacturer or an equivalent entity.

Each repaired sling shall be proof tested by the manufacturer or equivalent entity to twice the rated capacity prior to its return to service. The employer shall retain a certificate of the proof test and make it available for examination.

Slings, including webbing and fittings, which have been repaired in a temporary manner shall not be used.

Removal from service.
Synthetic web slings shall be immediately removed from service if any of the following conditions are present:

Acid or caustic burns;

Melting or charring of any part of the sling surface;

Snags, punctures, tears or cuts;

Broken or worn stitches; or

Distortion of fittings.

Interpretations.
Slings and all fastenings and attachments must be inspected for damage or defects each day before being used by a competent person designated by employer. Where service conditions warrant, additional inspections must be performed during sling use. Damaged or defective slings must be immediately removed from service.

As can be seen by the summary of the standards, there are many recordkeeping requirements for compliance. In many cases, each sling must be numbered and tagged, with a record kept of its use, repair history, date of inspection, etc. Because of the required detail, it is usually best to allow the stores personnel to track this information. Few, if any, maintenance departments dedicate the resources necessary to maintain the records necessary for compliance.

Hand and Power Tools

This section provides insight into employer responsibilities for the standards. In many of the regulations, it is the employer who is responsible for the condition of the tools, even if the tools are the property of the employees. Consider the following:

General Requirements.
Each employer shall be responsible for the safe condition of tools and equipment used by employees, including tools and equipment which may be furnished by employees.

Compressed air used for cleaning.
Compressed air shall not be used for cleaning purposes except where reduced to less than 30 p.s.i. and then only with effective chip guarding and personal protective equipment.

Cracked saws.
All cracked saws shall be removed from service.

Grounding.
Portable electric powered tools shall meet the electrical requirements of subpart S of this part. (proper grounding requirements)

Pneumatic Powered Tools and Hose.

Pneumatic power tools present some unique safety issues. The regulations address them specifically. The following is a sample of the specific regulations.

Tool retainer.
A tool retainer shall be installed on each piece of utilization equipment which, without such a retainer, may eject the tool.

Airhose.
Hose and hose connections used for conducting compressed air to utilization equipment shall be designed for the pressure and service to which they are subjected

Mounting and inspection of abrasive wheels for pneumatic grinders.
Immediately before mounting, all wheels shall be closely inspected and sounded by the user (ring test, see Subpart O, 1910.215(d)(1)) to make sure they have not been damaged in transit, storage, or otherwise. The spindle speed of the machine shall be checked before mounting of the wheel to be certain that it does not exceed the maximum operating speed marked on the wheel.

Grinding wheels shall fit freely on the spindle and remain free under all grinding conditions. A controlled clearance between the wheel hole and the machine spindle (or wheel sleeves or adaptors) is essential to avoid excessive pressure from mounting and spindle expansion. To accomplish this, the machine spindle shall be made to nominal (standard) size plus zero minus .002 inch, and the wheel hole shall be made suitably oversize to assure safety clearance under the conditions of operating heat and pressure.

All contact surfaces of wheels, blotters, and flangers shall be flat and free of foreign matter.

When a bushing is used in the wheel hole it shall not exceed the width of the wheel and shall not contact the flanges.

Explosive Power Tools.

The tool shall be so designed as not to be operable other than against a work surface, and unless the operator is holding the tool against the work surface with a force at least 5 pounds greater than the total weight of the tool.

No tools shall be loaded unless being prepared for immediate use, nor shall an unattended tool be left loaded.

Jacks.

All jacks shall be properly lubricated at regular intervals.

Each jack shall be thoroughly inspected at times which depend upon the service conditions. Inspections shall be not less frequent than the following:

For constant or intermittent use at one locality, once every 6 months,

For jacks sent out of shop for special work, when sent out and when returned,

For a jack subjected to abnormal load or shock, immediately before and immediately thereafter.

Repair or replacement parts shall be examined for possible defects.

Jacks which are out of order shall be tagged accordingly, and shall not be used until repairs are made.

Ladders.

The employer shall ensure that no employee nor any material or equipment may be supported or permitted to be supported on any portion of a ladder unless it is first determined, by inspections and checks conducted by a competent person that such ladder is adequately strong, in good condition, and properly secured in place, as required in Subpart D of this part and as required in this section.

The spacing between steps or rungs permanently installed on poles and towers shall be no more than 18 inches (36 inches on any one side).

This requirement also applies to fixed ladders on towers, when towers are so equipped. Spacing between steps shall be uniform above the initial unstepped section, except where working, standing, or access steps are required.

Fixed ladder rungs and step rungs for poles and towers shall have a minimum diameter of 5/8". Fixed ladder rungs shall have a minimum clear width of 12 inches. Steps for poles and towers shall have a minimum clear width of 4 1/2 inches. The spacing between detachable steps may not exceed 30 inches on any one side, and these steps shall be properly secured when in use.

Portable wood ladders intended for general use may not be painted but may be coated with a translucent nonconductive coating. Portable wood ladders may not be longitudinally reinforced with metal.

Portable wood ladders that are not being carried on vehicles and are not in active use shall be stored where they will not be exposed to the elements and where there is good ventilation.

These provisions shall apply to rolling ladders used in telecommunications centers, except that such ladders shall have a minimum inside width, between the side rails, of at least eight inches.

General.

The employer shall ensure that visual inspections are made of the equipment by a competent person each day the equipment is to be used to ascertain that it is in good condition.

As can be seen by the regulations, there is a tremendous need to inspect, service, and track the tools used by maintenance personnel. The stores function should be properly organized to help meet these requirements.

Summary

The inventory and purchasing functions play a large role in guaranteeing regulatory compliance. Best inventory practices insure that the proper parts are always on hand, not too many, but not too few. The inventory department also has the responsibility for seeing that the spares are properly stored and maintained prior to use. In addition, the stores personnel can be utilized to maintain and track the maintenance tools and equipment properly. In many cases, the stores personnel have been cross trained to make repairs on the maintenance tools and equipment. In this way, they help achieve regulatory compliance.

CHAPTER 3

Work Order Management Systems

The work order is the fundamental document to request, track, and record maintenance activities. It is also the key document utilized to plan and schedule maintenance activities. Using the work order, the planner will document all tools, equipment, spare parts, etc. that are required to perform the maintenance activities. Regulatory agencies place many requirements on the tools, equipment and even instructions that must be provided to the employees performing the work. This chapter explores many of the requirements and how the work order management system can help a company comply with the regulations.

Objectives for Work Order Management Systems

The work order or work management component of a maintenance strategy focuses on accomplishing work in the most safe and efficient manner. It also requires recording the work activities. This section reviews the regulations that address how maintenance work is performed, and includes procedural requirements from the regulatory agencies.

Scaffolding

This section could have been included in the stores chapter. However, the setup and dismantling of scaffolding often violates regulations; hence, it is included in the work management chapter.

Scaffolds are typically utilized in maintenance and overhaul projects where ladders or other work platforms are not sufficient to gain access to the area or equipment being serviced. By far, the majority of OSHA's fines levied during the 2000-2001 inspection year (greater than $13.5M) were related to scaffolding and fall protection violations. While the majority of these occurred in the construction trade, much can be learned and applied to the maintenance use of scaffolds. For example, maintenance work tends to be more short term than construction work. Therefore, the temptation exists to take shortcuts in setup. This temptation is even stronger because the scaffold only facilitates the work, it is not the work activity itself. Hence, the following common sense regulation from OSHA:

> The footing or anchorage for scaffolds shall be sound, rigid, and capable of carrying the maximum intended load without settling or displacement. Unstable objects such as barrels, boxes, loose brick, or concrete blocks shall not be used to support scaffolds or planks.

While one might reason this approach to building a scaffold would never be used, there are numerous violations of this standard. Therefore, all employees need to be aware of the regulations affecting the use of scaffolding. Common regulations about scaffolding include the following:

> Scaffolds and their components shall be capable of supporting without failure at least four times the maximum intended load.

> Scaffolds and other devices mentioned or described in this section shall be maintained in safe condition. Scaffolds shall not be altered or moved horizontally while they are in use or occupied.

> Scaffold planks shall extend over their end supports not less than 6 inches nor more than 18 inches.

> The poles, legs, or uprights of scaffolds shall be plumb, and securely and rigidly braced to prevent swaying and displacement

> Tools, materials, and debris shall not be allowed to accumulate in quantities to cause a hazard.

> All pole scaffolds shall be securely guyed or tied to the building or structure. Where the height or length exceeds 25 feet, the scaffold shall be secured at intervals not greater than 25 feet vertically and horizontally.

> Guardrails not less than 2 x 4 inches or the equivalent and not

less than 36 inches or more than 42 inches high, with a midrail, when required, of 1 x 4-inch lumber or equivalent, and toeboards, shall be installed at all open sides on all scaffolds more than 10 feet above the ground or floor. Toeboards shall be a minimum of 4 inches in height. Wire mesh shall be installed in accordance with paragraph (a)(17) of this section

Periodic inspections shall be made of all welded frames and accessories, and any maintenance, including painting, or minor corrections authorized by the manufacturer, shall be made before further use.

This regulation also includes the term "periodic inspections". These inspections help insure the structure of the scaffolding and its related equipment will be safe. Unfortunately, no one definite timetable for these inspections exists. As noted earlier in this book, the timing should be based on usage and conditions.

Hazard Assessment and Equipment Selection

Proper development of any job plan requires evaluation of the hazards that exist. It is then the responsibility of the employer to provide the necessary equipment, tools and personal protective equipment to insure the employees are safe while performing the work. There are many details that must be considered when specifying the proper safety equipment. These are detailed in the following section.

Personal Protective Equipment

The employer shall assess the workplace to determine if hazards are present, or are likely to be present, which necessitate the use of personal protective equipment (PPE). If such hazards are present, or likely to be present, the employer shall:

Select, and have each affected employee use, the types of PPE that will protect the affected employee from the hazards identified in the hazard assessment;

Communicate selection decisions to each affected employee; and,

Select PPE that properly fits each affected employee.

The employer shall verify that the required workplace hazard assessment has been performed through a written certi fication that identifies the workplace evaluated; the person certifying that the evaluation has been performed; the date(s) of the hazard assessment; and, which identifies the document as a certification of hazard assessment.

Defective and damaged equipment.
Defective or damaged personal protective equipment shall not be used.

PPE Training.

The employer shall provide training to each employee who is required by this section to use PPE. Each such employee shall be trained to know at least the following:

When PPE is necessary;

What PPE is necessary;

How to properly don, doff, adjust, and wear PPE;

The limitations of the PPE; and,

The proper care, maintenance, useful life and disposal of the PPE.

Each affected employee shall demonstrate an understanding of the training specified in this section, and the ability to use PPE properly, before being allowed to perform work requiring the use of PPE.

When the employer has reason to believe that any affected employee who has already been trained does not have the understanding and skill required by this section, the employer shall retrain each such employee. Circumstances where retraining is required include, but are not limited to, situations where:

Changes in the workplace render previous training obsolete; or

Changes in the types of PPE to be used render previous training obsolete; or

Inadequacies in an affected employee's knowledge or use of assigned PPE indicate that the employee has not retained the requisite understanding or skill.

The employer shall verify that each affected employee has received and understood the required training through a written certification that contains the name of each employee trained, the date(s) of training, and that identifies the subject of the certification.

When performing work in an area that requires PPE, the work order must contain the detail necessary to perform the work safely. In addition, it is essential that the supervisor insure that the employees assigned to the particular work activity have the proper training.

Confined Space Entry

This section looks at the requirements for performing work in areas that are deemed unsafe to enter. The following procedures are designed to insure employees are provided the knowledge and support (tools, training, etc) to perform the work in a safe manner.

"Hazardous atmosphere" means an atmosphere that may expose employees to the risk of death, incapacitation, impairment of ability to self-rescue (that is, escape unaided from a permit space), injury, or acute illness from one or more of the following causes:

(1) Flammable gas, vapor, or mist in excess of 10 percent of its lower flammable limit (LFL);

(2) Airborne combustible dust at a concentration that meets or exceeds its LFL;

NOTE: This concentration may be approximated as a condition in which the dust obscures vision at a distance of 5 feet (1.52 m) or less.

(3) Atmospheric oxygen concentration below 19.5 percent or above 23.5 percent;

(4) Atmospheric concentration of any substance for which a dose or a permissible exposure limit is published in Subpart G, Occupational Health and Environmental Control, or in

Subpart Z, Toxic and Hazardous Substances, of this Part and which could result in employee exposure in excess of its dose or permissible exposure limit;

NOTE: An atmospheric concentration of any substance that is not capable of causing death, incapacitation, impairment of ability to self-rescue, injury, or acute illness due to its health effects is not covered by this provision.

(5) Any other atmospheric condition that is immediately dangerous to life or health.

NOTE: For air contaminants for which OSHA has not determined a dose or permissible exposure limit, other sources of information, such as Material Safety Data Sheets that comply with the Hazard Communication Standard, section 1910.1200 of this Part, published information, and internal documents can provide guidance in establishing acceptable atmospheric conditions.

"Hot work permit" means the employer's written authoriza tion to perform operations (for example, riveting, welding, cutting, burning, and heating) capable of providing a source of ignition.

"Immediately dangerous to life or health (IDLH)" means any condition that poses an immediate or delayed threat to life or that would cause irreversible adverse health effects or that would interfere with an individual's ability to escape unaided from a permit space.

NOTE: Some materials — hydrogen fluoride gas and cadmium vapor, for example — may produce immediate transient effects that, even if severe, may pass without medical attention, but are followed by sudden, possibly fatal collapse 12-72 hours after exposure. The victim "feels normal" from recovery from transient effects until collapse. Such materials in hazardous quantities are considered to be "immediately" dangerous to life or health.

If the workplace contains permit spaces, the employer shall inform exposed employees, by posting danger signs or by any other equally effective means, of the existence and location of and the danger posed by the permit spaces.

NOTE: Sign reading DANGER — PERMIT-REQUIRED CONFINED SPACE, DO NOT ENTER or using other similar language would satisfy the requirement for a sign.

The employer shall verify that the space is safe for entry andthat the pre-entry measures required by this section have been taken, through a written certification that contains the date, the location of the space, and the signature of the person providing the certification. The certification shall be made before entry and shall be made available to each employee entering the space or to that employee's authorized representative .

The details of the documentation required before work is started are important to insure that all work environments are thoroughly checked in a timely fashion.

Training.

The employer shall provide training so that all employees whose work is regulated by this section acquire the understanding, knowledge, and skills necessary for the safe performance of the duties assigned under this section.

Training shall be provided to each affected employee:

Before the employee is first assigned duties under this section;

Before there is a change in assigned duties;

Whenever there is a change in permit space operations that presents a hazard about which an employee has not previously been trained;

Whenever the employer has reason to believe either that there are deviations from the permit space entry procedures required by paragraph (d)(3) of this section or that there are inadequacies in the employee's knowledge or use of these procedures.

The training shall establish employee ***proficiency*** in the duties required by this section and shall introduce new or revised procedures, as necessary, for compliance with this section.

The employer shall ***certify*** that the training required by this section has been accomplished. The certification shall contain each employee's name, the signatures or initials of the trainers, and the dates of training. The certification shall be available for inspection by employees and their authorized representatives.

This section is important because it covers the training for employees who actually perform the work activities. Proper training is especially important when operators or contractors work on the activities as well. For those without familiarity of the equipment and hazards, training is often the only thing that can prevent an accident.

Duties of authorized entrants.

The employer shall ensure that all authorized entrants:

Know the hazards that may be faced during entry, including information on the mode, signs or symptoms, and consequences of the exposure;

Properly use equipment as required by paragraph (d)(4) of this section;

Communicate with the attendant as necessary to enable the attendant to monitor entrant status and to enable the attendant to alert entrants of the need to evacuate the space as required by paragraph (i)(6) of this section;

Alert the attendant whenever:

> The entrant recognizes any warning sign or symptom of exposure to a dangerous situation, or
>
> The entrant detects a prohibited condition; and

Exit from the permit space as quickly as possible whenever:

> An order to evacuate is given by the attendant or the entry supervisor,
>
> The entrant recognizes any warning sign or symptom of exposure to a dangerous situation,
>
> The entrant detects a prohibited condition, or
>
> An evacuation alarm is activated.

Duties of attendants.

The employer shall ensure that each attendant:

Knows the hazards that may be faced during entry, including information on the mode, signs or symptoms, and consequences of the exposure;

Is aware of possible behavioral effects of hazard exposure in authorized entrants;

Continuously maintains an accurate count of authorized entrants in the permit space and ensures that the means used to identify authorized entrants under this section accurately identifies who is in the permit space;

Remains outside the permit space during entry operations until relieved by another attendant;

NOTE: When the employer's permit entry program allows attendant entry for rescue, attendants may enter a permit space to attempt a rescue if they have been trained and equipped for rescue operations as required this section and if they have been relieved as required by this section.

Communicates with authorized entrants as necessary to monitor entrant status and to alert entrants of the need to evacuate the space

Monitors activities inside and outside the space to determine if it is safe for entrants to remain in the space and orders the authorized entrants to evacuate the permit space immediately under any of the following conditions;

> If the attendant detects a prohibited condition;
>
> If the attendant detects the behavioral effects of hazard exposure in an authorized entrant;
>
> If the attendant detects a situation outside the space that could endanger the authorized entrants; or
>
> If the attendant cannot effectively and safely perform all the duties required under paragraph (i) of this section;

Summon rescue and other emergency services as soon as the attendant determines that authorized entrants may need assistance to escape from permit space hazards;

Takes the following actions when unauthorized persons approach or enter a permit space while entry is underway:

Warn the unauthorized persons that they must stay away from the permit space;

Advise the unauthorized pesons that they must exit immediately if they have entered the permit space; and

Inform the authorized entrants and the entry supervisor if unauthorized persons have entered the permit space;

Performs non-entry rescues as specified by the employer's rescue procedure; and

Performs no duties that might interfere with the atten dant's primary duty to monitor and protect the autho rized entrants.

Beyond just providing an attendant, many companies are required to provide additional training for the employees, based on the types of hazards that are encountered. These additional requirements are detailed in the following section.

An employer whose employees have been designated to provide permit space rescue and emergency services shall take the following measures:

Provide affected employees with the personal protective equipment (PPE) needed to conduct permit space rescues safely and train affected employees so they are proficient in the use of that PPE, at no cost to those employees;

Train affected employees to perform assigned rescue duties. The employer must ensure that such employees successfully complete the training required to establish proficiency as an authorized entrant, as provided by paragraphs (g) and (h) of this section;

Train affected employees in basic first-aid and cardiopulmonary resuscitation (CPR). The employer shall ensure that at least one member of the rescue team or service holding a current certification in first aid and CPR is available; and

Ensure that affected employees practice making permit space rescues at least once every 12 months, by means of simulated rescue operations in which they remove dummies, manikins, or actual persons from the actual permit spaces or from representative permit spaces. Representative permit spaces shall, with respect to opening size, configuration, and accessibility, simulate the types of permit spaces from which rescue is to be performed.

Lock Out /Tag Out

Almost every maintenance activity requires a knowledge of the Lock Out/Tag Out (LOTO) procedures. Therefore, the standards for LOTO are especially important for maintenance technicians. Many companies actually print the LOTO instructions and reminders on every work order that requires the procedures. This insures that maintenance technicians are provided the actual instructions. The following are guidelines from the OSHA regulations detailing LOTO regulations.

This standard applies to the control of energy during servicing and/or maintenance of machines and equipment.

Normal production operations are not covered by this standard. Servicing and/or maintenance which takes place during normal production operations is covered by this standard only if:

An employee is required to remove or bypass a guard or other safety device; or

An employee is required to place any part of his or her body into an area on a machine or piece of equipment where work is actually performed upon the material being processed (point of operation) or where an associated danger zone exists during a machine operating cycle.

Note: ***Exception:*** *Minor tool changes and adjustments, and other minor servicing activities, which take place during normal production operations, are not covered by this standard if they are routine, repetitive, and integral to the use of the equip-*

ment for production, provided that the work is performed using alternative measures which provide effective protection.

This standard does not apply to the following:

Work on cord and plug connected electric equipment for which exposure to the hazards of unexpected energization or start up of the equipment is controlled by the unplugging of the equipment from the energy source and by the plug being under the exclusive control of the employee performing the servicing or maintenance.

Energy control program.

The employer shall establish a program consisting of energy control procedures, employee training and periodic inspections to ensure that before any employee performs any servicing or maintenance on a machine or equipment where the unexpected energizing, startup or release of stored energy could occur and cause injury, the machine or equipment shall be isolated from the energy source and rendered inoperative

Lockout/ Tag out.

If an energy isolating device is not capable of being locked out, the employer's energy control program shall utilize a tag out system.

If an energy isolating device is capable of being locked out, the employer's energy control program shall utilize lockout, unless the employer can demonstrate that the utilization of a tag out system will provide full employee protection as set forth in this section.

After January 2, 1990, whenever replacement or major repair, renovation or modification of a machine or equipment is performed, and whenever new machines or equipment are installed, energy isolating devices for such machine or equipment shall be designed to accept a lockout device.

Full employee protection.

When a tag out device is used on an energy isolating device which is capable of being locked out, the tag out device shall be attached at the same location that the lockout device would have been attached, and the employer shall demonstrate that the tag out program will provide a level of safety equivalent to that obtained by using a lockout program.

In demonstrating that a level of safety is achieved in the tag out program which is equivalent to the level of safety obtained by using a lockout program, the employer shall demonstrate full compliance with all tagout-related provisions of this standard together with such additional elements as are necessary to provide the equivalent safety available from the use of a lockout device. Additional means to be considered as part of the demonstration of full employee protection shall include the implementation of additional safety measures such as the removal of an isolating circuit element, blocking of a controlling switch, opening of an extra disconnecting device, or the removal of a valve handle to reduce the likelihood of inadvertent energization.

Energy control procedure.

Procedures shall be developed, documented and utilized for the control of potentially hazardous energy when employees are engaged in the activities covered by this section.

Note: ***Exception****: The employer need not document the required procedure for a particular machine or equipment, when all of the following elements exist: (1) The machine or equipment has no potential for stored or residual energy or reaccumulation of stored energy after shut down which could endanger employees; (2) the machine or equipment has a single energy source which can be readily identified and isolated; (3) the isolation and locking out of that energy source will completely deenergize and deactivate the machine or equipment; (4) the machine or equipment is isolated from that energy source and locked out during servicing or maintenance; (5) a single lockout device will achieve a locker-out condition; (6) the lockout device is under the exclusive control of the authorized employee performing the servicing or maintenance; (7) the servicing or maintenance does not create hazards for other employees; and (8) the employer, in utilizing this exception, has had no accidents involving the unexpected activation or reenergization of the machine or equipment during servicing or maintenance.*

The procedures shall clearly and specifically outline the scope, purpose, authorization, rules, and techniques to be utilized for the control of hazardous energy, and the means to enforce compliance including, but not limited to, the following:

A specific statement of the intended use of the procedure;

Specific procedural steps for shutting down, isolating, blocking and securing machines or equipment to control hazardous energy;

Specific procedural steps for the placement, removal and transfer of lockout devices or tagout devices and the responsibility for them; and

Specific requirements for testing a machine or equipment to determine and verify the effectiveness of lockout devices, tagout devices, and other energy control measures

Protective materials and hardware.

Locks, tags, chains, wedges, key blocks, adapter pins, self-locking fasteners, or other hardware shall be provided by the employer for isolating, securing or blocking of machines or equipment from energy sources.

Lockout devices and tagout devices shall be singularly identified; shall be the only devices(s) used for controlling energy; shall not be used for other purposes; and shall meet the following requirements:

Durable.

Lockout and tagout devices shall be capable of withstanding the environment to which they are exposed for the maximum period of time that exposure is expected.

Tagout devices shall be constructed and printed so that exposure to weather conditions or wet and damp locations will not cause the tag to deteriorate or the message on the tag to become illegible.

Tags shall not deteriorate when used in corrosive environments such as areas where acid and alkali chemicals are handled and stored.

Standardized.

Lockout and tagout devices shall be standardized within the facility in at least one of the following criteria: Color; shape; or size; and additionally, in the case of tagout devices, print and format shall be standardized.

Substantial

Lockout devices.

Lockout devices shall be substantial enough to prevent removal without the use of excessive force or unusual techniques, such as with the use of bolt cutters or other metal cutting tools.

Tagout devices.

Tagout devices, including their means of attachment, shall be substantial enough to prevent inadvertent or accidental removal. Tagout device attachment means shall be of a non-reusable type, attachable by hand, self-locking, and non-releasable with a minimum unlocking strength of no less than 50 pounds and having the general design and basic characteristics of being at least equivalent to a one-piece, all environment-tolerant nylon cable tie.

Identifiable.

Lockout devices and tagout devices shall indicate the iden tity of the employee applying the device(s).

Tagout devices shall warn against hazardous conditions if the machine or equipment is energized and shall include a legend such as the following:

Do Not Start.

Do Not Open.

Do Not Close.

Do Not Energize.

Do Not Operate.

Periodic inspection.

The employer shall conduct a periodic inspection of the energy control procedure at least annually to ensure that the procedure and the requirements of this standard are being followed.

The periodic inspection shall be performed by an authorized employee other than the ones(s) utilizing the energy control procedure being inspected.

The periodic inspection shall be conducted to correct any deviations or inadequacies identified.

Where lockout is used for energy control, the periodic inspection shall include a review, between the inspector and each authorized employee, of that employee's responsibilities under the energy control procedure being inspected.

Where tagout is used for energy control, the periodic inspection shall include a review, between the inspector and each authorized and affected employee, of that employee's responsibilities under the energy control procedure being inspected, and the elements set forth in this section.

The employer shall certify that the periodic inspections have been performed. The certification shall identify the machine or equipment on which the energy control procedure was being utilized, the date of the inspection, the employees included in the inspection, and the person performing the inspection.

Training and communication.

The employer shall provide training to ensure that the purpose and function of the energy control program are understood by employees and that the knowledge and skills required for the safe application, usage, and removal of the energy controls are acquired by employees. The training shall include the following:

Each authorized employee shall receive training in the recognition of applicable hazardous energy sources, the type and magnitude of the energy available in the workplace, and the methods and means necessary for energy isolation and control.

Each affected employee shall be instructed in the purpose and use of the energy control procedure.

All other employees whose work operations are or may be in an area where energy control procedures may be utilized, shall be instructed about the procedure, and about the prohibition relating to attempts to restart or reenergize machines or equipment which are locked out or tagged out.

When tagout systems are used, employees shall also be trained in the following limitations of tags:

Tags are essentially warning devices affixed to energy isolating devices, and do not provide the physical restraint on those devices that is provided by a lock.

When a tag is attached to an energy isolating means, it is not to be removed without authorization of the authorized person responsible for it, and it is never to be bypassed, ignored, or otherwise defeated.

Tags must be legible and understandable by all authorized employees, affected employees, and all other employees whose work operations are or may be in the area, in order to be effective.

Tags and their means of attachment must be made of materials which will withstand the environmental conditions encountered in the workplace.

Tags may evoke a false sense of security, and their meaning needs to be understood as part of the overall energy control program.

Tags must be securely attached to energy isolating devices so that they cannot be inadvertently or accidentally detached during use.

Employee retraining.
Retraining shall be provided for all authorized and affected employees whenever there is a change in their job assignments, a change in machines, equipment or processes that present a

new hazard, or when there is a change in the energy control procedures.

Additional retraining shall also be conducted whenever a periodic inspection under paragraph (c)(6) of this section reveals, or whenever the employer has reason to believe that there are deviations from or inadequacies in the employee's knowledge or use of the energy control procedures.

The retraining shall reestablish employee proficiency and introduce new or revised control methods and procedures, as necessary.

The employer shall certify that employee training has been accomplished and is being kept up to date. The certification shall contain each employee's name and dates of training.

Energy isolation.
Lockout or tagout shall be performed only by the authorized employees who are performing the servicing or maintenance.

Notification of employees.
Affected employees shall be notified by the employer or authorized employee of the application and removal of lockout devices or tagout devices. Notification shall be given before the controls are applied, and after they are removed from the machine or equipment.

Stored energy.
Following the application of lockout or tagout devices to energy isolating devices, all potentially hazardous stored or residual energy shall be relieved, disconnected, restrained, and otherwise rendered safe.

If there is a possibility of reaccumulation of stored energy to a hazardous level, verification of isolation shall be continued until the servicing or maintenance is completed, or until the possibility of such accumulation no longer exists.

Verification of isolation.
Prior to starting work on machines or equipment that have been locked out or tagged out, the authorized employee shall verify that isolation and deenergization of the machine or equipment have been accomplished.

Release from lockout or tagout.
Before lockout or tagout devices are removed and energy is restored to the machine or equipment, procedures shall be followed and actions taken by the authorized employee(s) to ensure the following:

The machine or equipment.
The work area shall be inspected to ensure that nonessential items have been removed and to ensure that machine or equipment components are operationally intact.

Employees.
The work area shall be checked to ensure that all employees have been safely positioned or removed.

After lockout or tagout devices have been removed and before a machine or equipment is started, affected employees shall be notified that the lockout or tagout device(s) have been removed.

Lockout or tagout devices removal.
Each lockout or tagout device shall be removed from each energy isolating device by the employee who applied the device.

Exception: When the authorized employee who applied the lockout or tagout device is not available to remove it, that device may be removed under the direction of the employer, provided that specific procedures and training for such removal have been developed, documented and incorporated into the employer's energy control program. The employer shall demonstrate that the specific procedure provides equivalent safety to the removal of the device by the authorized employee who applied it. The specific procedure shall include at least the following elements:

Verification by the employer that the authorized employee who applied the device is not at the facility:

Making all reasonable efforts to contact the authorized employee to inform him/her that his/her lockout or tagout device has been removed; and

Ensuring that the authorized employee has this knowledge before he/she resumes work at that facility.

Group lockout or tagout.

When servicing and/or maintenance is performed by a crew, craft, department or other group, they shall utilize a procedure which affords the employees a level of protection equivalent to that provided by the implementation of a personal lockout or tagout device.

Group lockout or tagout devices shall be used in accordance with the procedures required by paragraph (c)(4) of this section including, but not necessarily limited to, the following specific requirements:

Primary responsibility is vested in an authorized employee for a set number of employees working under the protection of a group lockout or tagout device (such as an operations lock);

Provision for the authorized employee to ascertain the exposure status of individual group members with regard to the lockout or tagout of the machine or equipment and

When more than one crew, craft, department, etc. is involved, assignment of overall job-associated lockout or tagout control responsibility to an authorized employee designated to coordinate affected work forces and ensure continuity of protection; and

Each authorized employee shall affix a personal lockout or tagout device to the group lockout device, group lockbox, or comparable mechanism when he or she begins work, and shall remove those devices when he or she stops working on the machine or equipment being serviced or maintained.

Shift or personnel changes.

Specific procedures shall be utilized during shift or personel changes to ensure the continuity of lockout or tagout pro tection, including provision for the orderly transfer of lock out or tagout device protection between off-going and oncoming employees, to minimize exposure to hazards from the unexpected energization or start-up of the machine or equipment, or the release of stored energy.

Fire Watch

In performing various types of maintenance activities, combustible materials are often utilized. In addition, maintenance activities often require welding or cutting tools to be utilized. These situations, in addition to others create the potential for combustion. When these conditions are present, it is necessary to plan for a fire watch. The requirements for the fire watch are detailed in the following section.

> Fire watchers shall be required whenever welding or cutting is performed in locations where other than a minor fire might develop, or any of the following conditions exist:
>
> > Appreciable combustible material, in building construc tion or contents, closer than 35 feet (10.7 m) to the point of operation.
> >
> > Appreciable combustibles are more than 35 feet (10.7 m) away but are easily ignited by sparks.
> >
> > Wall or floor openings within a 35-foot (10.7 m) radius expose combustible material in adjacent areas including concealed spaces in walls or floors.
> >
> > Combustible materials are adjacent to the opposite side of metal partitions, walls, ceilings, or roofs and are likely to be ignited by conduction or radiation.
>
> Fire watchers shall have fire extinguishing equipment readily available and be trained in its use. They shall be familiar with facilities for sounding an alarm in the event of a fire. They shall watch for fires in all exposed areas, try to extinguish them only when obviously within the capacity of the equipment available, or otherwise sound the alarm. A fire watch shall be maintained for <u>at least a half hour after completion of welding or cutting operations</u> to detect and extinguish possible smoldering fires.

Management.

> Management shall recognize its responsibility for the safe usage of cutting and welding equipment on its property and:

Based on fire potentials of plant facilities, establish areas for cutting and welding, and establish procedures for cutting and welding, in other areas.

Designate an individual responsible for authorizing cutting and welding operations in areas not specifically designed for such processes.

Insist that cutters or welders and their supervisors are suitably trained in the safe operation of their equipment and the safe use of the process.

Advise all contractors about flammable materials or hazardous conditions of which they may not be aware.

Supervisor.

The Supervisor:

Shall be responsible for the safe handling of the cutting or welding equipment and the safe use of the cutting or welding process.

Shall determine the combustible materials and hazardous areas present or likely to be present in the work location.

Shall protect combustibles from ignition by the following:

Have the work moved to a location free from dangerous combustibles.

If the work cannot be moved, have the combustibles moved to a safe distance from the work or have the com bustibles properly shielded against ignition.

See that cutting and welding are so scheduled that plant operations that might expose combustibles to ignition are not started during cutting or welding.

Shall secure authorization for the cutting or welding operations from the designated management representative.

Shall determine that the cutter or welder secures his approval that conditions are safe before going ahead.

Shall determine that fire protection and extinguishing equipment are properly located at the site.

Where fire watches are required, he shall see that they are available at the site.

Protection from arc welding rays.

Where the work permits, the welder should be enclosed in an individual booth painted with a finish of low reflectivity such as zinc oxide (an important factor for absorbing ultraviolet radiations) and lamp black, or shall be enclosed with noncombustible screens similarly painted. Booths and screens shall permit circulation of air at floor level. Workers or other persons adjacent to the welding areas shall be protected from the rays by noncombustible or flameproof screens or shields or shall be required to wear appropriate goggles.

Warning sign.

After welding operations are completed, the welder shall mark the hot metal or provide some other means of warning other workers.

First-aid equipment.

First-aid equipment shall be available at all times. All injuries shall be reported as soon as possible for medical attention. First aid shall be rendered until medical attention can be provided.

Maximum pressure.

Under no condition shall acetylene be generated, piped (except in approved cylinder manifolds) or utilized at a pressure in excess of 15 psig (103 kPa gauge pressure) or 30 psia (206 kPa absolute). (The 30 psia (206 kPa absolute) limit is intended to prevent unsafe use of acetylene in pressurized chambers such as caissons, underground excavations or tunnel construction.) This requirement is not intended to apply to storage of acetylene dissolved in a suitable solvent in cylinders manufactured and maintained according to U.S. Department of Transport-ation requirements, or to acetylene for chemical use. The use of liquid acetylene shall be prohibited.

Personnel.

Workmen in charge of the oxygen or fuel-gas supply equipment, including generators, and oxygen or fuel-gas distribution piping systems shall be instructed and judged competent by their employers for this important work before being left in charge. Rules and instructions covering the operation and maintenance of oxygen or fuel-gas supply equipment including generators, and oxygen or fuel-gas distribution piping systems shall be readily available.

Compressed gas cylinders shall be legibly marked, for the purpose of identifying the gas content, with either the chemical or the trade name of the gas. Such marking shall be by means of stenciling, stamping, or labeling, and shall not be readily removable. Whenever practical, the marking shall be located on the shoulder of the cylinder. This method conforms to the American National Standard Method for Marking Portable Compressed Gas Containers to Identify the Material Contained, ANSI Z48.1-1954.

Compressed gas cylinders shall be equipped with connections complying with the American National Standard Compressed Gas Cylinder Valve Outlet and Inlet Connections, ANSI B57.1-1965.

All cylinders with a water weight capacity of over 30 pounds (13.6 kg) shall be equipped with means of connecting a valve protection cap or with a collar or recess to protect the valve.

Storage of cylinders-general

Cylinders shall be kept away from radiators and other sources of heat.

Inside of buildings, cylinders shall be stored in a well-protected, well-ventilated, dry location, at least 20 (6.1 m) feet from highly combustible materials such as oil or excelsior. Cylinders should be stored in definitely assigned places away from elevators, stairs, or gangways. Assigned storage spaces shall be located where cylinders will not be knocked over or damaged by passing or falling objects, or subject to tampering by unauthorized persons. Cylinders shall not be kept in unventilated enclosures such as lockers and cupboards.

Valve protection caps, where cylinder is designed to accept a cap, shall always be in place, hand-tight, except when cylinders are in use or connected for use.

Fuel gas cylinders

Fuel-gas cylinder storage. Inside a building, cylinders, except those in actual use or attached ready for use, shall be limited to a total gas capacity of 2,000 cubic feet (56 m(3)) or 300 pounds (135.9 kg) of liquefied petroleum gas.

Oxygen cylinders

Oxygen cylinders in storage shall be separated from fuel-gas cylinders or combustible materials (especially oil or grease), a minimum distance of 20 feet (6.1 m) or by a noncombustible barrier at least 5 feet (1.5 m) high having a fire-resistance rating of at least one-half hour.

Regulators

Before connecting a regulator to a cylinder valve, the valve shall be opened slightly and closed immediately. The valve shall be opened while standing to one side of the outlet; never in front of it. Never crack a fuel-gas cylinder valve near other welding work or near sparks, flame, or other possible sources of ignition.

Before a regulator is removed from a cylinder valve, the cylinder valve shall be closed and the gas released from the regulator.

An acetylene cylinder valve shall not be opened more than one and one-half turns of the spindle, and preferably no more than three-fourths of a turn.

Hose and hose connections.

Hose for oxy-fuel gas service shall comply with the Specification for Rubber Welding Hose, 1958, Compressed Gas Association and Rubber Manufacturers Association, which is incorporated by reference as specified in Sec. 1910.6.

When parallel lengths of oxygen and acetylene hose are taped together for convenience and to prevent tangling, not more than 4 inches (10.2 cm) out of 12 inches (30.5 cm) shall be covered by tape

Hose showing leaks, burns, worn places, or other defects rendering it unfit for service shall be repaired or replaced.

When regulators or parts of regulators, including gages, need repair, the work shall be performed by skilled mechanics who have been properly instructed.

Gages on oxygen regulators shall be marked "USE NO OIL."

Union nuts and connections on regulators shall be inspected before use to detect faulty seats which may cause leakage of gas when the regulators are attached to the cylinder valves.

Instruction.
Workmen designated to operate arc welding equipment shall have been properly instructed and qualified to operate such equipment as specified in paragraph (d) of this section.

Conduits containing electrical conductors shall not be used for completing a work-lead circuit. Pipelines shall not be used as a permanent part of a work-lead circuit, but may be used during construction, extension or repair providing current is not carried through threaded joints, flanged bolted joints, or caulked joints and that special precautions are used to avoid sparking at connection of the work-lead cable.

Chains, wire ropes, cranes, hoists, and elevators shall not be used to carry welding current.

Where a structure, conveyor, or fixture is regularly employed as a welding current return circuit, joints shall be bonded or provided with adequate current collecting devices.

All ground connections shall be checked to determine that they are mechanically strong and electrically adequate for the required current.

Operation and Maintenance

General.
Workmen assigned to operate or maintain arc welding equipment shall be acquainted with the requirements of this section and with 1910.252 (a), (b), and (c) of this part; if doing

gas-shielded arc welding, also Recommended Safe Practices for Gas-Shielded Arc Welding, A6.1-1966, American Welding Society, which is incorporated by reference as specified in Sec. 1910.6.

Machine hook up.
Before starting operations all connections to the machine shall be checked to make certain they are properly made. The work lead shall be firmly attached to the work; magnetic work clamps shall be freed from adherent metal particles of spatter on contact surfaces. Coiled welding cable shall be spread out before use to avoid serious overheating and damage to insulation.

Grounding.
Grounding of the welding machine frame shall be checked. Special attention shall be given to safety ground connections of portable machines.

Leaks.
There shall be no leaks of cooling water, shielding gas or engine fuel.

Switches.
It shall be determined that proper switching equipment for shutting down the machine is provided.

Manufacturers' instructions.
Printed rules and instructions covering operation of equipment supplied by the manufacturers shall be strictly followed.

Electrode holders.
Electrode holders when not in use shall be so placed that they cannot make electrical contact with persons, conducting objects, fuel or compressed gas tanks.

Electric shock.
Cables with splices within 10 feet (3 m) of the holder shall not be used. The welder should not coil or loop welding electrode cable around parts of his body.

Maintenance.
The operator should report any equipment defect or safety hazard to his supervisor and the use of the equipment shall be

discontinued until its safety has been assured. Repairs shall be made only by qualified personnel.

Machines which have become wet shall be thoroughly dried and tested before being used.

Cables with damaged insulation or exposed bare conductors shall be replaced. Joining lengths of work and electrode cables shall be done by the use of connecting means specifically intended for the purpose. The connecting means shall have insulation adequate for the service conditions.

Personnel.
Workmen designated to operate resistance welding equipment shall have been properly instructed and judged competent to operate such equipment.

Maintenance.
Periodic inspection shall be made by qualified maintenance personnel, and a certification record maintained. The certification record shall include the date of inspection, the signature of the person who performed the inspection and the serial number, or other identifier, for the equipment inspected. The operator shall be instructed to report any equipment defects to his supervisor and the use of the equipment shall be discontinued until safety repairs have been completed.

Employers should not allow employees to use compressed air for cleaning themselves or their clothing in general industry situations. The eyes and other body parts, such as the respiratory system, may be damaged as the result of inadequate personal protective equipment, lack of chip guards, and/or uncontrolled release of compressed air.

OSHA has no specification standards regulating the size, shape, and dimensions of hand tools. However, there is the general requirement that the tools and equipment used by employees be in a safe condition. In addition, there are some specific requirements relevant to “modified or mutilated” hand tools.

Battery handling.

In performing maintenance activities, it is necessary to work with batteries. This may be in servicing backup electrical systems or servicing batteries used to power material handling equipment. The regulations require careful planning to insure the safety of employees while working around batteries. The following section details some of these requirements.

> Eye protection devices which provide side as well as frontal eye protection for employees shall be provided when measuring storage battery specific gravity or handling electrolyte, and the employer shall ensure that such devices are used by the employees. The employer shall also ensure that acid resistant gloves and aprons shall be worn for protection against spattering. Facilities for quick drenching or flushing of the eyes and body shall be provided unless the storage batteries are of the enclosed type and equipped with explosion proof vents, in which case sealed water rinse or neutralizing packs may be substituted for the quick drenching or flushing facilities. Employees assigned to work with storage batteries shall be instructed in emergency procedures such as dealing with accidental acid spills.
>
> Electrolyte (acid or base, and distilled water) for battery cells shall be mixed in a well ventilated room. Acid or base shall be poured gradually, while stirring, into the water. Water shall never be poured into concentrated (greater than 75 percent) acid solutions. Electrolyte shall never be placed in metal containers nor stirred with metal objects.
>
> When taking specific gravity readings, the open end of the hydrometer shall be covered with an acid resistant material while moving it from cell to cell to avoid splashing or throwing the electrolyte.
>
> **Medical and first aid.**
>
> First aid supplies recommended by a consulting physician shall be placed in weatherproof containers (unless stored indoors) and shall be easily accessible. Each first aid kit shall be inspected at least once a month. Expended items shall be replaced.

Training.

Employers shall provide training in the various precautions and safe practices described in this section and shall insure that employees do not engage in the activities to which this section applies until such employees have received proper training in the various precautions and safe practices required by this section. However, where the employer can demonstrate that an employee is already trained in the precautions and safe practices required by this section prior to his employment, training need not be provided to that employee in accordance with this section. Where training is required, it shall consist of on-the-job training or classroom-type training or a combination of both. The employer shall certify that employees have been trained by preparing a certification record which includes the identity of the person trained, the signature of the employer or the person who conducted the training, and the date the training was completed. The certification record shall be prepared at the completion of training and shall be maintained on file for the duration of the employee's employment. The certification record shall be made available upon request to the Assistant Secretary for Occupational Safety and Health. Such training shall, where appropriate, include the following subjects:

Recognition and avoidance of dangers relating to encounters with harmful substances and animal, insect, or plant life;

Procedures to be followed in emergency situations; and,

First aid training, including instruction in artificial respiration.

Operations and Maintenance

The following information relates to the maintenance requirements specified in the EPA's Superfund provisions. The Superfund is provided to clean up sites where environmental damage has occurred. Once the site is cleaned up, operational and maintenance procedures are designed to prevent any reoccurrence of the problem. The following excerpts from the standard help to detail how extensively the maintenance procedures are supported by the work order management system.

Operation and maintenance (O&M) activities protect the integrity of the selected remedy for a site. O&M measures are initiated by a State after the remedy has achieved the remedial action objectives and remediation goals outlined in the record of decision (ROD), and is determined to be operational and functional (O&F) based on State and Federal agreement. For Superfund-lead sites, remedies are considered O&F either one year after construction is complete or when the remedy is functioning properly and performing as designed — whichever is earlier. Remedies requiring O&M measures include landfill caps, gas collection systems, ground water extraction treatment, ground water monitoring, or surface water treatment.

Once the O&M period begins, the State or Potentially Responsible Party (PRP) is responsible for maintaining the effectiveness of the remedy. O&M monitoring includes four components:

- inspection;
- sampling and analysis;
- routine maintenance; and
- reporting.

O&M activities are usually required for sites where cleanup proceeded through landfill/capping activities, ground water activities, or through natural attenuation.

Facility Maintenance

As part of maintaining the physical site, an owner of an agricultural establishment may have to be aware of environmental requirements that are not necessarily tied to agricultural operations. These requirements, while not agricultural in nature, must still be complied with by the owner. They cover the following areas:

- Asbestos
- Chemical Use and Safety
- Lead-Based Paint
- Managing Electrical Equipment With Polychlorinated Biphenyls (PCBs)
- Maintenance and Service of Unpaved Roads

Asbestos

Asbestos is the name for a group of naturally occurring minerals that separate into strong, very fine fibers. The fibers are heat-resistant and extremely durable. Because of these qualities, asbestos has become very useful in construction and industry. In the home it may or may not pose a health hazard to the occupants, depending on its condition. When it can be crushed by hand pressure or the surface is not sealed to prevent small pieces from escaping, the material is considered friable — fragile or easily crumbled. In this condition fibers can be released and pose a health risk. However, as long as the surface is stable and well-sealed against the release of its fibers and not damaged, the material is considered safe until damaged in some way.

Asbestos tends to break down into a dust of microscopic fibers. Because of their size and shape, these tiny fibers remain suspended in the air for long periods of time and can easily penetrate body tissues after being inhaled or ingested. Because of their durability, these fibers can remain in the body for many years and thereby become the cause of asbestos-related diseases. Symptoms of these diseases generally do not appear for 10 to 30 years after the exposure. Therefore, long before its effects are detectable, asbestos-related injury to the body may have already occurred. There is no safe level of exposure known; therefore, exposure to friable asbestos should be avoided.

Buildings on agricultural establishments and agribusinesses may contain asbestos or asbestos-containing materials (ACM). Used for insulation and as a fire retardant, asbestos and ACMs can be found in a variety of building construction materials, including pipe and furnace insulation materials, asbestos shingles, millboard, textured paint and other coating materials, and floor tiles. Asbestos may also be found in vehicle brakes. Buildings built in the sixties are more likely to have asbestos-containing sprayed- or troweled-on friable materials than other buildings.

Chemical Use and Safety

Whenever significant hazards are found in the course of accident investigations, EPA issues Chemical Safety Alerts to caution facilities, State Emergency Response Commissions, Local Emergency Planning Committees, emergency responders, and others to reduce risks and prevent future accidents.

Lead-Based Paint

In 1978, EPA banned the manufacture and use of lead-based paint and lead-based paint products. Current studies suggest that the primary sources of lead exposure for most children are deteriorating lead-based paint, lead-contaminated dust, and lead-contaminated residential soil. EPA is playing a major role in addressing these residential lead hazards. Lead-based paint chips and dust, if ingested, can create severe, long-term health effects, especially for children. Lead is a known carcinogen and, through any exposure pathway, may result in significant health effects.

Lead-based paint on an agricultural establishment or agribusiness farm will typically be found on building interiors and exteriors for buildings constructed prior to 1978. During renovation and demolition, paint removal has the potential to impact human health and the environment as fibers, dust, and paint chips are released. Paint chips and dust can cause indoor air contamination during renovation, and soil contamination from demolition or improper disposal.

Managing Electrical Equipment with Polychlorinated Biphenyls (PCBs)

PCBs are mixtures of synthetic organic chemicals that have the same basic chemical structure and that have physical properties ranging from oily liquids to waxy solids. Due to their non-flammability, chemical stability, high boiling point, and electrical insulating properties, PCBs were used in hundreds of industrial and commercial applications including electrical, heat transfer, and hydraulic equipment; as plasticizers in paints, plastics, and rubber products; in pigments, dyes, and carbonless copy paper; and in many other applications. More than 1.5 billion pounds of PCBs were manufactured in the United States before production stopped in 1977.

PCBs have significant ecological and human health effects including carcinogenicity (probable human cancer-causing or cancer-promoting agent), neurotoxicity, reproductive and developmental toxicity, immune system suppression, liver damage, skin irritation, and endocrine disruption. These toxic effects have been observed from both acute and chronic exposures to PCB mixtures with varying chlorine content. PCBs do not break down readily in the environment and are taken into the food chain by microorganisms. PCBs are then biologically accumulated and concentrated at levels much higher than found in the surrounding environment, thus posing a greater risk of injury to human health and the environment than might be imagined.

Polychlorinated biphenyls (PCBs) were widely used in electrical equipment manufactured from 1932 to 1978. Types of equipment on agricultural establishments and agribusinesses potentially containing PCBs include transformers and their bushings, capacitors, reclosers, regulators, electric light ballasts, and oil switches. Any pieces of equipment containing PCBs in their dielectric fluid at concentrations of greater than 50 ppm are subject to the PCB requirements. Human food or animal feed must not be exposed to PCBs. Therefore, transformers and other items containing PCBs must not be located near food or feed.

Maintenance and Service of Unpaved Roads

Erosion of unpaved roadways occurs when soil particles are loosened and carried away from the roadway base, ditch, or road bank by water, wind, traffic or other transport means. Exposed soils, high runoff velocities and volumes, sandy or silty soil types, and poor compaction increase the potential for erosion. Loosened soil particles are carried from the road bed and into the roadway drainage system. Some of these particles settle out satisfactorily in the road ditches, but most often they settle out where they diminished the carrying capacity of the ditch, and in turn cause roadway flooding, which subsequently leads to more roadway erosion. Most of the eroded soil, however, ultimately ends up in streams and rives where it diminishes channel capacity causing more frequent and severe flooding, destroys aquatic and riparian habitat, and has other adverse effects on water quality and water-related activities.

Chapter 3

EPA

The EPA requirements for maintenance documentation center on documenting the maintenance activities that are performed on environmental systems. This may range from preventive maintenance tracking to recording work on a work order and building the equipment's work order history. The following material presents a summation of the maintenance information requirements found in the EPA guidelines for the following operations:

- coating lines
- paint shops
- fabric coating
- paper coating
- vinyl coating
- coating of furniture
- large appliances
- wire coating
- miscellaneous parts coating
- wood panel coating

A maintenance log is to be maintained for the capture system, control device, and monitoring equipment detailing all routine and nonroutine maintenance performed including dates and duration of any outages.

Certification that each subject source at the facility is in compliance with the applicable emission limitation, equipment specification, or work practice.

The control device is equipped with the applicable monitoring equipment specified and the monitoring equipment is installed, calibrated, operated, and maintained according to the vendor's specifications at all times the control device is in use.

Gasoline loading–unloading facilities

Each calendar month, the vapor balance systems and each loading rack that loads gasoline tank trucks shall be inspected for liquid or vapor leaks during product transfer operations. For purposes of this paragraph, detection methods incorporating sight, sound, or smell are acceptable. Each leak that is detected shall be repaired within 15 calendar days after it is detected.

A record of each monthly leak inspection shall be kept on file at the plant. Inspection records shall include, at a minimum, the following information:

(i) Date of inspection.

(ii) Findings (may indicate no leaks discovered or location, nature, and severity of each leak).

(iii) Leak determination method.

(iv) Corrective action (date each leak repaired and reasons for any repair interval in excess of 15 calendar days).

(v) Inspector name and signature.

Recordkeeping and reporting requirements.

(1) The owner or operator of a gasoline tank truck subject to this section shall maintain records of all certification, testing, and repairs. The records shall identify the gasoline tank truck, the date of the tests or repair, and, if applicable, the type of repair and the date of retest. The records shall be maintained in a legible, readily available condition for at least 5 years after the date the testing or repair is completed. These records shall be made available to the Administrator immediately upon written or verbal request.

(2) The records required shall, at a minimum, contain:

(i) The gasoline tank truck vessel tank identification number.

(ii) The initial test pressure and the time of the reading.

(iii) The final test pressure and the time of the reading.

(iv) The initial test vacuum and the time of the reading.

(v) The final test vacuum and the time of the reading.

(vi) At the top of each report page, the company name and the date and location of the tests on that page.

(vii) The name and the title of person conducting the test.

Petroleum refining Equipment

Standards: Equipment inspection program.

(1) The owner or operator of a petroleum refinery shall conduct quarterly monitoring of each:

(i) Compressor.

(ii) Pump in light liquid service.

(iii) Valve in light liquid service,

(iv) Valve in gas/vapor service,

(v) Pressure relief valve in gas/vapor service,

(2) The owner or operator of a petroleum refinery shall conduct a weekly visual inspection of each pump in light liquid service.

(3) The owner or operator of a petroleum refinery shall monitor each pressure relief valve after each overpressure relief to ensure that the valve has properly reseated and is not leaking.

(4) When an instrument reading of 10,000 parts per million (ppm) or greater is measured, it shall be determined that a leak has been detected.

(5) If there are indications of liquid dripping from the equipment, it shall be determined that a leak has been detected.

(6) When a leak is detected, the owner or operator shall affix a weatherproof, readily visible tag in a bright color such as red or yellow bearing the equipment identification number and the date on which the leak was detected. This tag shall remain in place until the leaking equipment is repaired. The requirements of this paragraph apply to any leak detected by the equipment inspection program and to any leak from any equipment that is detected on the basis of sight, sound, or smell.

Standards: Alternative standards for valves—skip period leak detection and repair.

(1) An owner or operator shall comply initially with the requirements for valves in gas/vapor service and valves in light liquid service, as described in paragraph (d) of this section.

(2) After two consecutive quarterly leak detection periods with the percent of valves leaking equal or less than 2.0, an owner or operator may begin to skip one of the quarterly leak detection periods for the valves in gas/vapor and light liquid service.

(3) After five consecutive quarterly leak detection periods with the percent of valves leaking equal to or less than 2.0, an owner or operator may begin to skip 3 of the quarterly leak detection periods for the valves in gas/vapor and light liquid service.

(4) If the percent of valves leaking is greater than 2.0, the owner or operator shall comply with the requirements as described in paragraph (d) of this section but can again elect to use the requirements in paragraph (e) of this section.

(5) The percent of valves leaking shall be determined by dividing the sum of valves found leaking during current monitoring and valves for which repair has been delayed by the total number of valves subject to the requirements of this section.

(6) An owner or operator shall keep a record of the percent of valves found leaking during each leak detection period.

Standards: Alternative standards for unsafe-to-monitor valves and difficult-to-monitor valves.

(1) Any valve that is designated, as an unsafe-to-monitor valve is exempt from the requirements of paragraph (d) if:

(i) The owner or operator of the valve demonstrates that the valve is unsafe to monitor because monitoring personnel would be exposed to an immediate danger as a consequence of complying with paragraph (d).

(ii) The owner or operator of the valve adheres to a written plan that requires monitoring of the valve as frequently as practicable during safe-to-monitor times.

(2) Any valve that is designated, as described in paragraph (j)(5)(i), as a difficult-to-monitor valve is exempt from the requirements of paragraph (d) if:

(i) The owner or operator of the valve demonstrates that the valve cannot be monitored without elevating the monitoring personnel more than 2 meters (m) (6.6 feet [ft]) above a sup port surface.

(ii) The owner or operator of the valve follows a written plan that requires monitoring of the valve at least once per calendar year.

Standards: Equipment repair program.
The owner or operator of a petroleum refinery shall:

(1) Make a first attempt at repair for any leak not later than 5 calendar days after the leak is detected.

(2) Repair any leak as soon as practicable, but not later than 15 calendar days after it is detected

Standards: Delay of repair.
(1) Delay of repair of equipment for which a leak has been detected is allowed if the repair is technically infeasible without a process unit shutdown. Repair of such equipment shall occur before the end of the next process unit shutdown.

(2) Delay of repair of equipment is allowed for equipment that is isolated from the process and that does not remain in VOC service.

(3) Delay of repair beyond a process unit shutdown is allowed for a valve, if valve assembly replacement is necessary during the process unit shutdown, valve assembly supplies have been depleted, and valve assembly supplies had been sufficiently stocked before the supplies were depleted. Delay of repair beyond the next process unit shutdown is not allowed unless the next process unit shutdown occurs sooner than 6 months after the first process unit shutdown.

Recordkeeping requirements.
(1) Each owner or operator subject to the provisions of this section shall comply with the recordkeeping requirements of this

section. Except as noted, these records shall be maintained in a readily accessible location for a minimum of 5 years and shall be made available to the Administrator immediately upon verbal or written request.

(2) An owner or operator of more than one affected facility subject to the provisions of this section may comply with the recordkeeping requirements for these facilities in one recordkeeping system if the system identifies each record by each facility.

(3) When each leak is detected as specified in paragraph (d) of this section, the following information shall be recorded in a log and shall be kept for 5 years in a readily accessible location:

(i) The instrument and operator identification numbers and the equipment identification number.

(ii) The date the leak was detected and the dates of each attempt to repair the leak.

(iii) The repair methods employed in each attempt to repair the leak.

(iv) omitted

(v) The notation "Repair delayed" and the reason for the delay if a leak is not repaired within 15 calendar days after discovery of the leak.

(vi) The signature of the owner or operator (or designate) whose decision it was that repair could not be effected with out a process shutdown.

(vii) The expected date of successful repair of the leak if a leak is not repaired within 15 calendar days.

(viii) The dates of process unit shutdowns that occur while the equipment is unrepaired.

(ix) The date of successful repair of the leak.

(4) A list of identification numbers of equipment in vacuum service shall be recorded in a log that is kept in a readily accessible location.

(5) The following information pertaining to all valves subject to the requirements of paragraph (f) of this section shall be recorded in a log that is kept for 5 years in a readily accessible location:

(i) A list of identification numbers for valves that are desig nated as unsafe to monitor, an explanation for each valve stating why the valve is unsafe to monitor, and the plan for monitoring each valve.

(d) Standards: Equipment inspection program.
The owner or operator of a natural gas/gasoline processing facility subject to this section shall conduct the equipment inspection program

(1) The owner or operator of a natural gas/gasoline processing facility subject to this section shall conduct quarterly monitoring of each:

(i) Compressor.

(ii) Pump in light liquid service.

(iii) Valve in light liquid service, except as provided in paragraphs (e) and (f) of this section.

(iv) Valve in gas/vapor service, except as provided in paragraphs (e) and (f) of this section.

(v) Pressure relief valve in gas/vapor service, except as provided in paragraphs (e) and (f) of this section.

(2) The owner or operator of a natural gas/gasoline processing facility subject to this section shall conduct a weekly visual inspection of each pump in light liquid service.

(3) The owner or operator of a natural gas/gasoline processing facility subject to this section shall monitor each pressure relief valve within 5 days after each overpressure relief to ensure that the valve has properly reseated and is not leaking.

(4)(i) Any pressure relief device that is located in a nonfractionating plant that is monitored only by nonplant personnel may be monitored after a pressure release the next time the monitoring personnel are on site, instead of within 5 days.

Recordkeeping.
Each owner or operator of a solvent metal cleaning source subject to this section shall maintain the following records in a readily accessible location for at least 5 years and shall make these records available to the Administrator upon verbal or written request:

(1) A record of central equipment maintenance, such as replacement of the carbon in a carbon adsorption unit.

(2) The results of all tests conducted in accordance with the requirements in paragraph (d) of this section.

Reporting.
The owner or operator of any facility containing sources subject to this section shall:

(1) Comply with the initial compliance certification requirements of XX.3003(a) of this subpart.

(2) Comply with the requirements of XX.3003(b) of this subpart regarding reports of excess emissions.

(3) Comply with the requirements of XX.3003(c) of this subpart for excess emissions related to any control devices used to comply with paragraphs (c)(1)(iii)(C),

The owner or operator of a pharmaceutical manufacturing facility subject to this section shall maintain the following records:

(i) Parameters listed in paragraph (e) of this section shall be recorded.

(ii) For sources subject to this section, the solvent true vapor pressure as determined by ASTM D323-89 shall be recorded for every process.

(1) Omitted –

(2) For any leak subject to paragraph (c)(6) of this section, which cannot be readily repaired within 1 hour after detection, the following records shall be kept:

(i) The name of the leaking equipment.
(ii) The date and time the leak is detected.

(iii) The action taken to repair the leak.
(iv) The date and time the leak is repaired.

Printing Presses

Recordkeeping.

On and after {insert date 1 year after promulgation of final rule}, or on and after the initial startup date, the owner or operator of a printing press subject to the limitations of this section and complying by means of control devices shall collect and record all of the following information each day for each printing press and maintain the information at the facility for a period of 5 years:

(A) Control device monitoring data.

(B) A log of operating time for the capture system, con trol device, monitoring equipment and the associated print ing press.

(C) A maintenance log for the capture system, control device, and monitoring equipment detailing all routine and non-routine maintenance performed including dates and duration of any outages.

Manufacturing Facitlities

Standards: Equipment inspection program.

The owner or operator of a synthetic organic chemical, polymer, or resin manufacturing facility shall conduct the equipment inspection program described in paragraphs (d)(1) through (d)(3) of this section using the test methods specified in XX.3085 of this subpart. The owner or operator of a synthetic organic chemical, polymer, or resin manufacturing facility shall conduct quarterly monitoring of each:

(i) Compressor.

(ii) Pump in light liquid service.

(iii) Valve in light liquid service, except as provided in paragraphs (e) and (f) of this section.

(iv) Valve in gas/vapor service, except as provided in paragraphs (e) and (f) of this section.

(v) Pressure relief valve in gas/vapor service, except as provided in paragraphs (e) and (f) of this section.

The owner or operator of a synthetic organic chemical, polymer, or resin manufacturing facility shall monitor each pressure relief valve after each overpressure relief to ensure that the valve has properly reseated and is not leaking.

(i) When an instrument reading of 10,000 parts per mil lion (ppm) or greater is measured, it shall be determined that a leak has been detected.

(ii) If there are indications of liquid dripping from the equipment, it shall be determined that a leak has been detected.

When a leak is detected, the owner or operator shall affix a weatherproof, readily visible tag in a bright color such as red or yellow bearing the equipment identification number and the date on which the leak was detected. This tag shall remain in place until the leaking equipment is repaired. The requirements of this paragraph apply to any leak detected by the equipment inspection program and to any leak from any equipment that is detected on the basis of sight, sound, or smell.

Standards: Equipment repair program.
The owner or operator of a synthetic organic chemical, polymer, or resin manufacturing facility refinery shall:

(1) Make a first attempt at repair for any leak not later than 5 calendar days after the leak is detected.

(2) Repair any leak as soon as practicable, but not later than 15 calendar days after it is detected except as provided in paragraph (h) of this section.

Standards: Delay of repair.
(1) Delay of repair of equipment for which a leak has been detected is allowed if repair is technically infeasible without a process unit shutdown. Repair of such equipment shall occur before the end of the first process unit shutdown after the leak is detected.

(2) Delay of repair of equipment is also allowed for equipment that is isolated from the process and that does not remain in VOC service after the leak is detected.

(3) Delay of repair beyond a process unit shutdown is allowed for a valve, if valve assembly replacement is necessary during the process unit shutdown, and if valve assembly supplies have been depleted, where valve assembly supplies had been sufficiently stocked before the supplies were depleted. Delay of repair beyond the first process unit shutdown is not allowed unless the next process unit shutdown occurs sooner than 6 months after the first process unit shutdown.

When each leak is detected as specified in paragraph (d) of this section, the following information shall be recorded in a log and shall be kept for 5 years in a readily accessible location:

(i) The instrument and operator identification numbers and the equipment identification number.

(ii) The date the leak was detected and the dates of each attempt to repair the leak.

(iii) The repair methods employed in each attempt to repair the leak.

(iv) The notation "Above 10,000" if the maximum instrument reading measured by the methods specified in XX.3085 of this subpart after each repair attempt is equal to or greater than 10,000 ppm.

(v) The notation "Repair delayed" and the reason for the delay if a leak is not repaired within 15 calendar days after the leak is discovered.

(vi) The signature of the owner or operator (or designate) whose decision it was that repair could not be effected without a process shutdown.

(vii) The expected date of successful repair of the leak if a leak is not repaired within 15 days.

(viii) The dates of process unit shutdowns that occur

while the equipment is unrepaired.

(ix) The date of successful repair of the leak.

The owner or operator of an air oxidation facility that uses a boiler or process heater to seek to comply with paragraph (c)(1) of this section shall install, calibrate, maintain, and operate according to the manufacturer's specifications the following equipment:

(i) A flow indicator that provides a record of vent stream flow to the boiler or process heater at least once every hour for each air oxidation facility. The flow indicator shall be installed in the vent stream from each air oxidation reactor within a facility at a point closest to the inlet of each boiler or process heater and before being joined with any other vent stream.

(ii) A temperature monitoring device in the firebox equipped with a continuous recorder and having an accuracy of ±1 percent of the temperature being measured expressed in degrees Celsius or ±0.5 °C, whichever is greater, for boilers or process heaters of less than 44 megawatts (MW) (150 million British thermal units per hour [Btu/hr]) heat input design capacity.

(iii) Monitor and record the periods of operation of the boiler or process heater if the design input capacity of the boiler or process heater is 44 MW (150 million Btu/hr) or greater. The records shall be readily avail able for inspection.

The owner or operator of an air oxidation facility that seeks to demonstrate compliance with the TRE index value limit specified under paragraph (c)(3) of this section shall install, calibrate, maintain, and operate according to manufacturer's specifications the following equipment:

(i) Where an absorber is the final recovery device in a recovery system:

(A) A scrubbing liquid temperature monitoring device having an accuracy of ±1 percent of the temperature being monitored, expressed in degrees Celsius or

±0.5°C, whichever is greater, and a specific gravity monitoring device having an accuracy of ±0.02 specific gravity unit, each equipped with a continuous recorder.

(B) An organic monitoring device used to indicate the concentration level of organic compounds exiting the recovery device based on a detection principle such as infrared, photoionization, or thermal conductivity, each equipped with a continuous recorder.

Continuous Emission Monitoring Systems (CEMS)

(a) CEMS quality control (QC) program. Each owner or operator of a CEMS shall develop and implement a CEMS QC pro gram. At a minimum, each QC program shall include written procedures that describe in detail step-by-step procedures and operations for each of the following:

(1) Initial and routine periodic calibration of the CEMS.

(2) Calibration drift (CD) determination and adjustment of the CEMS.

(3) Preventative maintenance of the CEMS (including spare parts inventory).

(4) Data recording, calculations, and reporting.

(5) Accuracy audit procedures including sampling and analysis methods.

(6) Program of corrective action for malfunctioning CEMS.

Food and Drug Administration

The Food and Drug Administration span of control far exceeds plant and facilities maintenance requirements. However, the maintenance requirements are fundamental for any plant or facility to achieve and then maintain FDA approval. Work order systems are once again the key tracking document for all maintenance activities. It will be impossible to maintain compliance without accurate work order records as the following material highlights.

Facilities and Equipment System

This system includes the measures and activities which provide an appropriate physical environment and resources used in the production of the drugs or drug products. It includes:

a) Buildings and facilities along with maintenance;

Equipment qualifications (installation and operation); equipment calibration and preventative maintenance; and cleaning and validation of cleaning processes as appropriate. Process performance qualification will be evaluated as part of the inspection of the overall process validation which is done within the system where the process is employed; and,

Utilities that are not intended to be incorporated into the product such as HVAC, compressed gases, steam and water systems.

For each of the following, the firm should have written and approved procedures and documentation resulting therefrom. The firm's adherence to written procedures should be verified through observation whenever possible. These areas may indicate deficiencies not only in this system but also in other systems that would warrant expansion of coverage. When this system is selected for coverage in addition to the Quality System, all areas listed below should be covered; however, the depth of coverage may vary depending upon inspectional findings.

Facilities.

Cleaning and maintenance

facility layout and air handling systems for prevention of cross-contamination (e.g. penicillin, beta-lactams, steroids, hormones, cytotoxics, etc.)

specifically designed areas for the manufacturing operations performed by the firm to prevent contamination or mix-ups

general air handling systems

control system for implementing changes in the building

lighting, potable water, washing and toilet facilities, sewage and refuse disposal

sanitation of the building, use of rodenticides, fungicides, insecticides, cleaning and sanitizing agents

Equipment.
The following are some additional equipment requirements.

equipment installation and operational qualification where appropriate

adequacy of equipment design, size, and location

equipment surfaces should not be reactive, additive, or absorptive

appropriate use of equipment operations substances, (lubricants, coolants, refrigerants, etc.) contacting products/containers/etc.

cleaning procedures and cleaning validation

controls to prevent contamination, particularly with any pesticides or any other toxic materials, or other drug or non-drug chemicals

qualification, calibration and maintenance of storage equipment, such as refrigerators and freezers for ensuring that standards, raw materials, reagents, etc. are stored at the proper temperatures

equipment qualification, calibration and maintenance, including computer qualification/validation and security

control system for implementing changes in the equipment

equipment identification practices (where appropriate)

documented investigation into any unexpected discrepancy

In this chapter, many of the recordkeeping requirements for regulatory compliance were considered. Many of the requirements were quite extensive, requiring substantial details for the plant and facility equipment. Since this chapter dealt with the work order management system, it can easily be seen that regulatory compliance is impossible without a fully utilized system.

It is up to each company to provide, not only the work order management system, but the resources to properly utilize the system. This is a great challenge for companies in the present business environment of constantly reducing headcount.

One tool that has been useful in enhancing the utilization of the work order system is the computerized maintenance management system which is considered in the next chapter.

CHAPTER 4

Computerized Maintenance Management Systems

It is clear from the regulations considered to this point that the maintenance department is required to keep accurate records spanning everything from personnel to work management detail and equipment condition. Understand the requirements for a computerized record keeping system, and how such a system impacts the regulations, is important. While no regulation specifically requires computerized record keeping, the volume of data that is required to be maintained mandates that some form of a computerized database should be utilized. While all of the agencies require the recordkeeping (as was detailed in the previous chapter) this chapter will consider the following requirements from the FDA:

Food and Drug Administration

The following regulations require equipment numbering schemes to identify and track each specific piece of equipment. Also, if any equipment is able to be moved, an equipment location tracking scheme needs to be developed.

Equipment identification.

> Major equipment shall be identified by a distinctive identification number or code that shall be recorded in the batch production record to show the specific equipment used in the manufacture of each batch of a drug product. In cases where only one of a particular type of equipment exists in a manufacturing facility, the name of the equipment may be used in lieu of a distinctive identification number or code.

Regulations require the accurate tracking of work orders for all work performed on the equipment. Most CMMS do not include the production tracking data. However, by chronologically matching the work performed to the production log, all of the maintenance record history can be referenced in detail.

Equipment cleaning and use log.

> A written record of major equipment cleaning, maintenance (except routine maintenance such as lubrication and adjustments), and use shall be included in individual equipment logs that show the date, time, product, and lot number of each batch processed. If equipment is dedicated to manufacture of one product, then individual equipment logs are not required, provided that lots or batches of such product follow in numerical order and are manufactured in numerical sequence. In cases where dedicated equipment is employed, the records of cleaning, maintenance, and use shall be part of the batch record. The persons performing and double-checking the cleaning and maintenance shall date and sign or initial the log indicating that the work was performed. Entries in the log shall be in chronological order.

Another regulation requires that original documents be scanned and stored electronically. Most CMMS allow the electronic attachment of scanned documents.

> Records required under this part may be retained either as original records or as true copies such as photocopies, microfilm, microfiche, or other accurate reproductions of the original records. Where reduction techniques, such as microfilming, are used, suitable reader and photocopying equipment shall be readily available.

The next regulation for laboratories clearly shows the need for a work order tracking log. This regulation can be met by combining the work order and PM history.

Laboratory records.

> Complete records shall be maintained of the periodic calibration of laboratory instruments, apparatus, gauges, and recording devices.

The next regulation covers the records that must be kept for site inspections. The items that follow are on the FDA inspector's checklist.

Facilities and Equipment System.

For each of the following, the firm should have written and approved procedures and documentation resulting therefrom. The firm's adherence to written procedures should be verified through observation whenever possible. These areas may indicate deficiencies not only in this system but also in other systems that would warrant expansion of coverage. When this system is selected for coverage in addition to the Quality System, all areas listed below should be covered; however, the depth of coverage may vary depending upon inspectional findings.

1. Facilities

— cleaning and maintenance

— facility layout and air handling systems for prevention of cross-contamination (e.g. penicillin, beta-lactams, steroids, hormones, cytotoxics, etc.)

— specifically designed areas for the manufacturing operations performed by the firm to prevent contamination or mix-ups

— general air handling systems

— control system for implementing changes in the building

2. Equipment

— equipment installation and operational qualification where appropriate

— adequacy of equipment design, size, and location

— equipment surfaces should not be reactive, additive, or absorptive

— appropriate use of equipment operations substances, (lubricants, coolants, refrigerants, etc.) contacting products / containers/etc.

- — cleaning procedures and cleaning validation
- — controls to prevent contamination, particularly with any pesticides or any other toxic materials, or other drug or non-drug chemicals

3. Maintenance supplies

- — qualification, calibration and maintenance of storage equipment, such as refrigerators and freezers for ensuring that standards, raw materials, reagents, etc. are stored at the proper temperatures
- — equipment qualification, calibration and maintenance, including computer qualification/validation and security
- — control system for implementing changes in the equipment
- — equipment identification practices (where appropriate)
- — documented investigation into any unexpected discrep ancy

4. Production System.

- — training/qualification of personnel
- — control system for implementing changes in processes
- — adequate procedure and practice for charge-in of components
- — identification of equipment with contents, and where appropriate phase of manufacturing and/or status
- — implementation and documentation of in-process controls, tests, and examinations (e.g., pH, adequacy of mix, weight variation, clarity)
- — justification and consistency of in-process specifications and drug product final specifications
- — prevention of objectionable microorganisms in non-sterile drug products
- — adherence to preprocessing procedures (e.g., set-up, line clearance, etc.)

- — equipment cleaning and use logs
- — master production and control records
- — batch production and control records
- — process validation, including validation and security of computerized or automated processes
- — change control; the need for revalidation evaluated
- — documented investigation into any unexpected discrepancy

Electronic signatures.

Amid the demand for considerable record keeping to meet compliance guidelines, there have been debates on the usage of electronic signatures. This issue is important for verification. The is the most stringent when it comes to electronic signatures. Other agencies accept electronic signatures more readily. Technology advancements have led to several methods of electronic record keeping that will satisfy even the most stringent inspector. Consider the following to see how the regulations can be met.

a) The regulations in this part set forth the criteria under which the agency considers electronic records, electronic signatures, and handwritten signatures executed to electronic records, to be trustworthy, reliable, and generally equivalent to paper records and handwritten signatures executed on paper.

b) This part applies to records in electronic form that are created, modified, maintained, archived, retrieved, or transmitted, under any records requirements set forth in agency regulations. This part also applies to electronic records submitted to the agency under requirements of the Federal Food, Drug, and Cosmetic Act and the Public Health Service Act, even if such records are not specifically identified in agency regulations. However, this part does not apply to paper records that are, or have been, transmitted by electronic means.

c) Where electronic signatures and their associated electronic records meet the requirements of this part, the agency will consider the electronic signatures to be equivalent to full handwritten signatures, initials, and other general signings as required by agency regulations, unless specifically excepted by regulation(s) effective on or after August 20, 1997.

d) Electronic records that meet the requirements of this part may be used in lieu of paper records unless paper records are specifically required.

e) Computer systems (including hardware and software), controls, and attendant documentation maintained under this part shall be readily available for, and subject to, FDA inspection.

Implementation.

a) For records required to be maintained, but not submitted to the agency, persons may use electronic records in lieu of paper records or electronic signatures in lieu of traditional signatures, in whole or in part, provided that the requirements of this part are met

b) For records submitted to the agency, persons may use electronic records in lieu of paper records or electronic signatures in lieu of traditional signatures, in whole or in part, provided that:

(1) The requirements of this part are met; and

(2) The document or parts of a document to be submitted have been identified in public docket No. 92S-0251 as being the type of submission the agency accepts in electronic form. This docket will identify specifically what types of documents or parts of documents are acceptable for submission in electronic form without paper records and the agency receiving unit(s) (e.g., specific center, office, division, branch) to which such submissions may be made. Documents to agency receiving unit(s) not specified in the public docket will not be considered as official if they are submitted in electronic form; paper forms of such documents will be considered as official and must accompany any electronic records. Persons are expected to consult with the intended agency receiving unit for details on how (e.g., method of transmission, media, file formats, and technical protocols) and whether to proceed with the electronic submission.

Definitions.

a) The definitions and interpretations of terms contained in section 201 of the act apply to those terms when used in this part.

b) The following definitions of terms also apply to this part:

(1) Act means the Federal Food, Drug, and Cosmetic Act (secs. 201-903 (21 U.S.C. 301-393)).

(2) Agency means the Food and Drug Administration.

(3) Biometrics means a method of verifying an individual's identity based on measurement of the individual's physical feature(s) or repeatable action(s) where those features and/or actions are both unique to that individual and measurable.

(4) Closed system means an environment in which system access is controlled by persons who are responsible for the content of electronic records that are on the system.

(5) Digital signature means an electronic signature based upon cryptographic methods of originator authentication, computed by using a set of rules and a set of parameters such that the identity of the signer and the integrity of the data can be verified.

(6) Electronic record means any combination of text, graph ics, data, audio, pictorial or other information representation in digital form, that is created, modified, maintained, archived, retrieved or distributed by a computer system.

(7) Electronic signature means a computer data compilation of any symbol or series of symbols, executed, adopted or authorized by an individual to be the legally binding equivalent of the individual's handwritten signature.

(8) Handwritten signature means the scripted name or legal mark of an individual, handwritten by that individual and executed or adopted with the present intention to authenticate a writing in a permanent form. The act of signing with a writing or marking instrument such as a pen or stylus is preserved. The scripted name or legal mark, while conventionally applied to paper, may also be applied to other devices that capture the name or mark.

(9) Open system means an environment in which system access is not controlled by persons who are responsible for the content of electronic records that are on the system.

Controls for closed systems.

Persons who use closed systems to create, modify, maintain, or transmit electronic records shall employ procedures and controls designed to ensure the authenticity, integrity, and, when appropriate, the confidentiality of electronic records, and to ensure that the signer cannot readily repudiate the signed record as not genuine. Such procedures and controls shall include the following:

a) Validation of systems to ensure accuracy, reliability, consistent intended performance, and the ability to discern invalid or altered records.

b) The ability to generate accurate and complete copies of records in both human readable and electronic form suitable for inspection, review, and copying by the agency. Persons should contact the agency if there are any questions regarding the ability of the agency to perform such review and copying of the electronic records

c) Protection of records to enable their accurate and ready retrieval throughout the records retention period.

d) Limiting system access to authorized individuals.

e) Use of secure, computer-generated, time-stamped audit trails to independently record the date and time of operator entries and actions that create, modify, or delete electronic records. Record changes shall not obscure previously recorded information. Such audit trail documentation shall be retained for a period at least as long as that required for the subject electronic records and shall be available for agency review and copying.

f) Use of operational system checks to enforce permitted sequencing of steps and events, as appropriate.

g) Use of authority checks to ensure that only authorized individuals can use the system, electronically sign a record, access the operation or computer system input or output device, alter a record, or perform the operation at hand.

h) Use of device (e.g., terminal) checks to determine, as appropriate, the validity of the source of data input or operational instruction.

i) Determination that persons who develop, maintain, or use electronic record/electronic signature systems have the education, training, and experience to perform their assigned tasks.

j) The establishment of, and adherence to, written policies that hold individuals accountable and responsible for actions initiated under their electronic signatures, in order to deter record and signature falsification.

Some additional consideration, for non-biometrical electronic signatures, would include the following:

a) Electronic signatures that are not based upon biometrics shall:

(1) Employ at least two distinct identification components such as an identification code and password.

(2) Be used only by their genuine owners; and

(3) Be administered and executed to ensure that attempted use of an individual's electronic signature by anyone other than its genuine owner requires collaboration of two or more individuals.

Persons who use electronic signatures based upon use of identification codes in combination with passwords shall employ controls to ensure their security and integrity. Such controls shall include:

a) Maintaining the uniqueness of each combined identification code and password, such that no two individuals have the same combination of identification code and password.

b) Ensuring that identification code and password issuances are periodically checked, recalled, or revised, (e.g., to cover such events as password aging).

c) Following loss management procedures to electronically deauthorize lost, stolen, missing, or otherwise potentially compromised tokens, cards, and other devices that bear or generate identification code or password information, and to issue temporary or permanent replacements using suitable, rigorous controls.

Electronic signatures for CVM

In another FDA document (How to Use E-Mail to Submit Information to The Center for Veterinary Medicine (CVM) Feb 1, 2001), the agency shows a greater willingness to use electronic format.. The following are requirements from that document.

> Each individual who intends to submit electronic submissions will have a unique electronic signature to verify the sender's identity. The unique electronic signature will consist of the sender's e-mail address and the individual password.
>
> This individual password should consist of 12 case-sensitive alphanumeric
>
> characters. The sender uses the individual password to encrypt the file before transmitting it to CVM. Only CVM's electronic submission system and the person who submits the electronic submission should know the individual password and be able to open the file.
>
> CVM will acknowledge receipt of the electronic submission by creating a new e-mail message (i.e., not by using a "Reply" feature) and sending it to the e-mail address of the individual submitting the electronic submission. The receipt will be a password-encrypted file of routing information from STARS attached to an e-mail message. CVM will use the sponsor password to encrypt the file.
>
> Individuals who submit information electronically must follow the electronic signature requirements described in 21 CFR 11.100-11.300. These requirements must be followed for CVM to accept electronic submissions as the official copy, instead of paper. If sponsors are not capable of meeting these requirements sponsors must submit information

on paper. With regard to the procedures set out in the final guidance that the sponsors should use to submit information electronically, on request CVM will discuss alternative methods of submitting the information.

Again, as technologies become more sophisticated, the regulatory agencies are allowing greater freedom in the use of electronic media for data storage, retrieval and transmission. Even further detail (from the FDA's perspective) is given in the following document. While the data requirements given here are for clinical trials, the acceptance of sign in/ password tracking is accepted.

A. Electronic Signatures for Data in Clinical Trials

The following material relates to clinical trials and starts with section V – Data Entry in the Clinical Trials Regulations

1. To ensure that individuals have the authority to proceed with data entry, the data entry system should be designed so that individuals need to enter electronic signatures, such as combined identification codes/passwords or biometric-based electronic signatures, at the start of a data entry session.

2. The data entry system should also be designed to ensure attributability. Therefore, each entry to an electronic record, including any change, should be made under the electronic signature of the individual making that entry. However, this does not necessarily mean a separate electronic signature for each entry or change. For example, a single electronic signature may cover multiple entries or changes.

   a. The printed name of the individual who enters data should be displayed by the data entry screen throughout the data entry session. This is intended to preclude the possibility of a different individual inadvertently entering data under someone else's name.

   b. If the name displayed by the screen during a data entry session is not that of the person entering the data, then that individual should log on under his or her own name before continuing.

The above regulations places some requirements on a computer system to insure that the above requirements are met. Any company purchasing an electronic system to track the FDA data would need to check the software thoroughly to insure the specifications are met.

3. Individuals should only work under their own passwords or other access keys and should not share these with others. Individuals should not log on to the system in order to provide another person access to the system.

4. Passwords or other access keys should be changed at estab lished intervals.

5. When someone leaves a workstation, the person should log off the system. Failing this, an automatic log off may be appropriate for long idle periods. For short periods of inactivity, there should be some kind of automatic protection against unatho rized data entry. An example could be an automatic screen saver that prevents data entry until a password is entered.

The following requirements set standards for audit tracking. These standards require a higher level of sophistication than the level found in many off-the-shelf software packages.

B. Audit Trails

1. Section 21 CFR 11.10(e) requires persons who use electronic record systems to maintain an audit trail as one of the procedures to protect the authenticity, integrity, and, when appropriate, the confidentiality of electronic records.

a. Persons must use secure, computer-generated, time-stamped audit trails to independently record the date and time of operator entries and actions that create, modify, or delete electronic records. A record is created when it is saved to durable media, as described under "commit" in Section II, Definitions.

b. Audit trails must be retained for a period at least as long as that required for the subject electronic records (e.g., the study data and records to which they pertain) and must be available for agency review and copying.

2. Personnel who create, modify, or delete electronic records should not be able to modify the audit trails.

3. Clinical investigators should retain either the original or a certified copy of audit trails.

4. FDA personnel should be able to read audit trails both at the study site and at any other location where associated electronic study records are maintained.

5. Audit trails should be created incrementally, in chronological order, and in a manner that does not allow new audit trail information to overwrite existing data in violation of 11.10(e).

C. Date/Time Stamps

1. Controls should be in place to ensure that the system's date and time are correct.

2. The ability to change the date or time should be limited to authorized personnel and such personnel should be notified if a system date or time discrepancy is detected. Changes to date or time should be documented.

3. Dates and times are to be local to the activity being documented and should include the year, month, day, hour, andminute. The Agency encourages establishments to synchronize systems to the date and time provided by trusted third parties.

4. Clinical study computerized systems will likely be used in multi-center trials, perhaps located in different time zones. Calculation of the local time stamp may be derived in such cases from a remote server located in a different time zone.

Electronic Signatures Training Requirements

Many organization do not concentrate on training their personnel extensively in the use of computer systems. The following regulation goes beyond initial training and requires additional training be given when software/ system upgrades are performed.

TRAINING OF PERSONNEL

A. Qualifications

1. Each person who enters or processes data should have the education, training, and experience or any combination thereof necessary to perform the assigned functions.

2. Individuals responsible for monitoring the trial should have education, training, and experience in the use of the computerized system necessary to adequately monitor the trial.

B. Training

1. Training should be provided to individuals in the specific operations that they are to perform.

2. Training should be conducted by qualified individuals on a continuing basis, as needed, to ensure familiarity with the computerized system and with any changes to the system during the course of the study.

C. Documentation

Employee education, training, and experience should be documented.

From this chapter, it is clear that electronic record keeping is acceptable in most regulatory agencies. However, the management and computer systems have specific requirements of the regulations are to be met. A disciplined approach to data collection is paramount. Fully utilizing the system, to insure accurate data is also important. Finally, training to insure proficiency in using the system is necessary for compliance. If these areas are properly addressed, electronic record keeping can do much to alleviate the work required to maintain compliance.

CHAPTER 5

Technical Training and Skills Development

Regulatory agencies require skilled professionals to audit procedures and perform various inspections. This chapter considers the perspective that OSHA applies to the topic of technical skills re-quirements. While the other regulatory agencies also require skilled professionals, the OSHA regulations are more specific. It can be safely assumed that the other agencies look for the same compliance that OSHA requires without specifically listing them in detail.

OSHA

The OSHA guidelines for inspections and service of maintenance tools and equipment required skilled, technical personnel. These requirements include Lock Out/ Tag Out (LOTO) and Personal Protective Equipment (PPE). If personnel are not previously qualified, then the employer is required to provide necessary training. The OSHA guidelines are even specific to the format of training. The requirements for OSHA begin with the definition of a "Competent Person".

Competent Persons

OSHA guidelines require a competent person to perform inspections and servicing of equipment. But who is considered competent? The following direct quotes are OSHA responses to letters requesting the definition of a competent person:

> There is wide latitude in the regulations as to the definition of a "competent person." A competent person can be the equipment

> owner's maintenance personnel or any other person the owner chooses, as long as that person is deemed "competent." A competent person does not have to be a disinterested third party.
>
> Section 29 CFR 1926.32 defines many of the terms used in the construction safety and health regulations including many used in the sections pertaining to crane safety. Subsection (b) defines a "competent person" as one who is capable of identifying existing and predictable hazards involved and is authorized to take prompt corrective measures to eliminate them. Subsection (h) defines a "designated person" as meaning an "authorized person" as defined in subsection (d). Subsection (d) defines an "authorized person" as one who the employer has approved or assigned to perform a specific type of duty or duties at specific locations at the job site.
>
> OSHA's position is that a person who does not have a thorough knowledge of the requirements, regulations and standards governing his/her direct duties cannot be considered as being a "competent person." This position has been upheld in a number of contested cases.

These quotes are of particular interest to the maintenance personnel who, for the most part, will be considered competent personnel. However, do operational or non-technical employees qualify as competent persons? The answer depends on the particular item being inspected or serviced. However, in the case of technical equipment inspections, the employee's skill set must include technical proficiency to comply with the standards. Consider the following requirements for training for electrical work.

> **Training.**
> Employees shall be trained in and familiar with the safety-related work practices, safety procedures, and other safety requirements in this section that pertain to their respective job assignments. Employees shall also be trained in and familiar with any other safety practices, including applicable emergency procedures (such as pole top and manhole rescue), that are not specifically addressed by this section but that are related to their work and are necessary for their safety.

Qualified employees shall also be trained and competent in:

The skills and techniques necessary to distinguish exposed live parts from other parts of electric equipment,

The skills and techniques necessary to determine the nominal voltage of exposed live parts,

The minimum approach distances specified in this section corresponding to the voltages to which the qualified employee will be exposed, and

The proper use of the special precautionary techniques, personal protective equipment, insulating and shielding materials, and insulated tools for working on or near exposed energized parts of electric equipment.

Note: For the purposes of this section, a person must have this training in order to be considered a qualified person.

The employer shall determine, through regular supervision and through inspections conducted on at least an annual basis, that each employee is complying with the safety-related work practices required by this section.

An employee shall receive additional training (or retraining) under any of the following conditions:

If the supervision and annual inspections required by paragraph(a)(2)(iii) of this section indicate that the employee is not complying with the safety-related work practices required by this section, or

If new technology, new types of equipment, or changes in procedures necessitate the use of safety-related work practices that are different from those which the employee would normally use, or

If he or she must employ safety-related work practices that are not normally used during his or her regular job duties.

Note: OSHA would consider tasks that are peformed less often than once per year to necessitate retraining before the performance of the work practices involved.

The training required by paragraph (a)(2) of this section shall be of the classroom or on-the-job type.

The training shall establish employee proficiency in the work practices required by this section and shall introduce the procedures necessary for compliance with this section.

The employer shall certify that each employee has received the training required by paragraph (a)(2) of this section. This certification shall be made when the employee demonstrates proficiency in the work practices involved and shall be maintained for the duration of the employee's employment.

Note: Employment records that indicate that an employee has received the required training are an acceptable means of meeting this requirement.

The following sections discuss the requirements for Lock Out/Tag Out training and some guidelines for refresher training. The requirements also specify additional training when job responsibilities are changed.

General.

The employer shall establish a program consisting of energy control procedures, employee training, and periodic inspections to ensure that, before any employee performs any servicing or maintenance on a machine or equipment where the unexpected energizing, start up, or release of stored energy could occur and cause injury, the machine or equipment is isolated from the energy source and rendered inoperative.

After November 1, 1994, whenever replacement or major repair, renovation, or modification of a machine or equipment is performed, and whenever new machines or equipment are installed, energy isolating devices for such machines or equipment shall be designed to accept a lockout device.

Procedures shall be developed, documented, and used for the control of potentially hazardous energy covered by paragraph (d) of this section.

Retraining shall be provided by the employer as follows:

> Retraining shall be provided for all authorized and affected employees whenever there is a change in their job assignments, a change in machines, equipment, or processes that present a new hazard or whenever there is a change in the energy control procedures.
>
> Retraining shall also be conducted whenever a periodic inspection under paragraph (d)(2)(v) of this section reveals, or whenever the employer has reason to believe, that there are deviations from or inadequacies in an employee's knowledge or use of the energy control procedures.
>
> The retraining shall reestablish employee proficiency and shall introduce new or revised control methods and procedures, as necessary.

The employer shall certify that employee training has been accomplished and is being kept up to date. The certification shall contain each employee's name and dates of training.

Individual Industry Requirements

Some training requirements in the OSHA guidelines are very specific ones applied to individual industries. Instead of presenting the training requirements for each industry, the training requirements for a specific industry — the grain industry — will be presented. Further application of the training requirements can then be made to each individual industry.

Training.

The employer shall provide training to employees at least annually and when changes in job assignment will expose them to new hazards. Current employees, and new employees prior to starting work, shall be trained in at least the following:

General safety precautions associated with the facility, including recognition and preventive measures for the hazards related to dust accumulations and common ignition sources such as smoking; and,

Specific procedures and safety practices applicable to their job tasks including but not limited to, cleaning procedures for grinding equipment, clearing procedures for choked legs, housekeeping procedures, hot work procedures, preventive maintenance procedures and lock-out/tag-out procedures.

Employees assigned special tasks, such as bin entry and handling of flammable or toxic substances, shall be provided training to perform these tasks safely.

Note: Training for an employee who enters grain storage structures includes training about engulfment and mechanical hazards and how to avoid them.

The employer shall develop and implement a written housekeeping program that establishes the frequency and method(s) determined best to reduce accumulations of fugitive grain dust on ledges, floors, equipment, and other exposed surfaces.

In addition, the housekeeping program for grain elevators shall address fugitive grain dust accumulations at priority housekeeping areas.

Priority housekeeping areas shall include at least the following:

Floor areas within 35 feet (10.7 m) of inside bucket elevators;
Floors of enclosed areas containing grinding equipment;
Floors of enclosed areas containing grain dryers located inside the facility.

The employer shall immediately remove any fugitive grain dust accumulations whenever they exceed 1/8 inch (.32 cm) at priority housekeeping areas, pursuant to the housekeeping program, or shall demonstrate and assure, through the development and implementation of the housekeeping program, that equivalent protection is provided

It is important that employees be trained in the recognition and prevention of hazards associated with grain facilities, especially those hazards associated with their own work tasks. Employees should understand the factors which are necessary to produce a fire or explosion, i.e., fuel (such as grain dust), oxygen, ignition source, and (in the case of explosions) confinement. Employees should be made aware that any efforts they make to keep these factors from occurring simultaneously will be an important step in reducing the potential for fires and explosions.

The standard provides flexibility for the employer to design a training program which fulfills the needs of a facility. The type, amount, and frequency of training will need to reflect the tasks that employees are expected to perform. Although training is to be provided to employees at least annually, it is recommended that safety meetings or discussions and drills be conducted at more frequent intervals.

The training program should include those topics applicable to the particular facility, as well as topics such as: Hot work procedures; lock-out/tag-out procedures; bin entry procedures; bin cleaning procedures; grain dust explosions; fire prevention; procedures for handling "hot grain"; housekeeping procedures, including methods and frequency of dust removal; pesticide and fumigant usage; proper use and maintenance of personal protective equipment; and, preventive maintenance. The types of work clothing should also be considered in the program at least to caution against using polyester clothing that easily melts and increases the severity of burns, as compared to wool or fire retardant cotton.

In implementing the training program, it is recommended that the employer utilize films, slide-tape presentations, pamphlets, and other information which can be obtained from such sources as the Grain Elevator and Processing Society, the Cooperative Extension Service of the U.S. Department of Agriculture, Kansas State University's Extension Grain Science and Industry, and other state agriculture schools, industry associations, union organizations, and insurance groups.

Standard Number: 1910.272 App A.

The standard also requires the employer to develop and implement procedures consisting of locking out and tagging equipment to prevent the inadvertent application of energy or motion to equipment being repaired, serviced, or adjusted, which could result in employee injury. All employees who have responsibility for repairing or servicing equipment, as well as those who operate the equipment, are to be familiar with the employer's lock and tag procedures. A lock is to be used as the positive means to prevent operation of the disconnected equipment. Tags are to be used to inform employees why equipment is locked out. Tags are to meet requirements in 1910.145(f). Locks and tags may only be removed by employees that placed them, or by their supervisor, to ensure the safety of the operation.

As can be seen, the training requirements are quite extensive for the grain industry. Almost all industries have similar regulations. Training is one of the most important aspects of keeping an employee safe. Yet, many companies ignore the need to train their employees. This area needs to be addressed more completely in all industrial settings today.

CHAPTER 6

Operations Involvement

This chapter focuses on regulations that cover areas in which maintenance and operations, and even other departments, may share responsibilities. These areas include general plant functions, and in particular the instances when the maintenance function has a direct impact. At the same time, the maintenance department is not solely responsible for compliance in these areas.

First Aid and Fire Protection

In many plants, the production process presents hazards to the workers. To insure rapid treatment in the event of an accident, companies provide medical care and even fire protection on the site. Where the possibility of injury exists and companies have their own emergency service, the regulations in the following sections apply.

Emergency Services

The following material applies to the services that are required to be provided to employees The requirements are so specific that they include response time based on the type of accident that is likely to occur in a given location.

Response time.
In areas where accidents resulting in suffocation, severe bleeding, or other life threatening or permanently disabling injury or illness can be expected, a 3 to 4 minute response time, from time of injury to time of administering first aid, is required. In other circumstances, i.e., where a life-threatening or permanently disabling injury is an unlikely outcome of an accident, a longer response time such as 15 minutes is acceptable.

> Where first aid treatment cannot be administered to injured employees by outside professionals within the required response time for the expected types of injuries, a person or persons within the facility shall be adequately trained to render first aid.

Maintenance personnel carry out their activities in remote locations. It is difficult for the company to insure that the first aid provisions can be complied with in a timely fashion. Radio or cell communications can help expedite reaching an injured employee in a remote location. Options should be explored to insure compliance with this provision.

Organizational statement.

> The employer shall prepare and maintain a statement or written policy which establishes the existence of
>
> a fire brigade;
>
> the basic organizational structure;
>
> the type, amount, and frequency of training to be provided to fire brigade members;
>
> the expected number of members in the fire brigade; and
>
> the functions that the fire brigade is to perform at the work place.

The organizational statement shall be available for inspection by the Assistant Secretary and by employees or their designated representatives.

Personnel.

The employer shall assure that employees who are expected to do interior structural fire fighting are physically capable of performing duties which may be assigned to them during emergencies. The employer shall not permit employees with known heart disease, epilepsy, or emphysema, to participate in fire brigade emergency activities unless a physician's certificate of the employees' fitness to participate in such activities is provided.

Training and education.

The employer shall provide training and education for all fire brigade members commensurate with those duties and functions that fire brigade members are expected to perform. Such training and education shall be provided to fire brigade members before they perform fire brigade emergency activities. Fire brigade leaders and training instructors shall be provided with training and education which is more comprehensive than that provided to the general membership of the fire brigade.

The employer shall assure that training and education is conducted frequently enough to assure that each member of the fire brigade is able to perform the member's assigned duties and functions satisfactorily and in a safe manner so as not to endanger fire brigade members or other employees. All fire brigade members shall be provided with training at least annually. In addition, fire brigade members who are expected to perform interior structural fire fighting shall be provided with an education session or training at least quarterly.

The employer shall inform fire brigade members about special hazards such as:

storage and use of flammable liquids and gases,

toxic chemicals,

radioactive sources, and

water reactive substances,

to which they may be exposed during fire and other emergencies. The fire brigade members shall also be advised of any changes that occur in relation to the special hazards. The employer shall develop and make available for inspection by fire brigade members, written procedures that describe the actions to be taken in situations involving the special hazards and shall include these in the training and education program.

Fire fighting equipment.

The next provision calls for good inspection of fire brigade equipment. A company that complies with this provision insures that the equipment is inspected as a regular part of a PM program.

The employer shall maintain and inspect, at least annually, fire fighting equipment to assure the safe operational condition of the equipment. Portable fire extinguishers and respirators shall be inspected at least monthly. Fire fighting equipment that is in damaged or unserviceable condition shall be removed from service and replaced.

Protective Clothing

The following requirements apply to those employees who perform interior structural fire fighting. The requirements do not apply to employees who use fire extinguishers or standpipe systems to control or extinguish fires only in the incipient stage.

General

The employer shall provide at no cost to the employee and assure the use of protective clothing which complies with the requirements of this paragraph. The employer shall assure that protective clothing ordered or purchased after July 1, 1981, meets the requirements contained in this paragraph. As the new equipment is provided, the employer shall assure that all fire brigade members wear the equipment when performing interior structural fire fighting. After July 1, 1985, the employer shall assure that all fire brigade members wear protective clothing meeting the requirements of this paragraph when performing interior structural fire fighting.

The employer shall assure that protective clothing protects the head, body, and extremities, and consists of at least the following components: foot and leg protection; hand protection; body protection; eye, face and head protection.

Foot and leg protection.

Foot and leg protection shall meet the requirements of the paragraphs of this section, and may be achieved by either of the following methods:

Fully extended boots which provide protection for the legs; or

Protective shoes or boots worn in combination with protective trousers that meet the requirements of paragraph (e)(3) of this section.

Protective footwear shall meet the requirements of 1910.136 for Class 75 footwear. In addition, protective footwear shall be water-resistant for at least 5 inches (12.7 cm) above the bottom of the heel and shall be equipped with slip-resistant outer soles.

Protective footwear shall be tested in accordance with paragraph (1) of Appendix E, and shall provide protection against penetration of the midsole by a size 8D common nail when at least 300 pounds (1330 N) of static force is applied to the nail.

Body protection.

Body protection shall be coordinated with foot and leg protection to ensure full body protection for the wearer. This shall be achieved by one of the following methods:

Wearing of a fire-resistive coat meeting the requirements of paragraph (e)(3)(ii) of this section in combination with fully extended boots meeting the requirements of paragraphs (e)(2)(ii) and (e)(2)(iii) of this section; or

Wearing of a fire-resistive coat in combination with protective trousers both of which meet the requirements of paragraph (e)(3)(ii) of this section.

The performance, construction, and testing of fire-resistive coats and protective trousers shall be at least equivalent to the requirements of the National Fire Protection Association (NFPA) standard NFPA No. 1971-1975, “Protective Clothing for Structural Fire Fighting,” which is incorporated by reference as specified in Sec. 1910.6, (See Appendix D to Subpart L) with the following permissible variations from those requirements:

Tearing strength of the outer shell shall be a minimum of 8 pounds (35.6 N) in any direction when tested in accordance with paragraph (2) of Appendix E; and

The outer shell may discolor but shall not separate or melt when placed in a forced air laboratory oven at a temperature of 500 deg. F (260 deg. C) for a period of five minutes. After cooling to ambient temperature and using the test method specified in paragraph (3) of Appendix E, char length shall not exceed 4.0 inches (10.2 cm) and after-flame shall not exceed 2.0 seconds.

Hand protection.

Hand protection shall consist of protective gloves or glove system which will provide protection against cut, puncture, and heat penetration. Gloves or glove system shall be tested in accordance with the test methods contained in the National Institute for Occupational Safety and Health (NIOSH) 1976 publication, "The Development of Criteria for Fire Fighter's Gloves; Vol. II, Part II: Test Methods," which is incorporated by reference as specified in Sec. 1910.6, (See Appendix D to Subpart L) and shall meet the following criteria for cut, puncture, and heat penetration:

Materials used for gloves shall resist surface cut by a blade with an edge having a 60 deg. included angle and a .001 inch (.0025 cm.) radius, under an applied force of 16 lbf (72N), and at a slicing velocity of greater or equal to 60 in/min (2.5 cm./sec);

Materials used for the palm and palm side of the fingers shall resist puncture by a penetrometer (simulating a 4d lath nail), under an applied force of 13.2 lbf (60N), and at a velocity greater or equal to 20 in/min (.85 cm./sec); and

The temperature inside the palm and gripping surface of the fingers of gloves shall not exceed 135 deg. F (57 deg. C) when gloves or glove system are exposed to 932 deg. F (500 deg. C) for five seconds at 4 psi (28 kPa) pressure.

Exterior materials of gloves shall be flame resistant and shall be tested in accordance with paragraph (3) of Appendix E. Maximum allowable afterflame shall be 2.0 seconds, and the maximum char length shall be 4.0 inches (10.2 cm).

When design of the fire-resistive coat does not otherwise provide protection for the wrists, protective gloves shall have wristlets of at least 4.0 inches (10.2 cm) in length to protect the wrist area when the arms are extended up-ward and outward from the body.

Head, eye and face protection.

Head protection shall consist of a protective head device with ear flaps and chin strap which meet the performance, construction, and testing requirements of the National Fire Safety and

Research Office of the National Fire Prevention and Control Administration, U.S. Department of Commerce (now known as the U.S. Fire Administration), which are contained in "Model Performance Criteria for Structural Firefighters' Helmets" (August 1977) which is incorporated by reference as specified in Sec. 1910.6, (See Appendix D to Subpart L).

Protective eye and face devices which comply with 1910.133 shall be used by fire brigade members when performing operations where the hazards of flying or falling materials which may cause eye and face injuries are present. Protective eye and face devices provided as accessories to protective head devices (face shields) are permitted when such devices meet the requirements of 1910.133.

Full facepieces, helmets, or hoods of breathing apparatus which meet the requirements of 1910.134 and paragraph (f) of this section, shall be acceptable as meeting the eye and face protection requirements of paragraph (e)(5)(ii) of this section.

Respiratory Protection.

This section covers the respirators that are used by the fire brigade.They must meet the same requirements as those mentioned earlier in the book.

General.

The employer must ensure that respirators are provided to, and used by, fire brigade members, and that the respirators meet the requirements of 29 CFR 1910.134 and this paragraph.

Approved self-contained breathing apparatus with full-facepiece, or with approved helmet or hood configuration, shall be provided to and worn by fire brigade members while working inside buildings or confined spaces where toxic products of combustion or an oxygen deficiency may be present.

Such apparatus shall also be worn during emergency situations involving toxic substances.

Approved self-contained breathing apparatus may be equipped with either a "buddy-breathing" device or a quick disconnect valve, even if these devices are not certified by NIOSH. If these accessories are used, they shall not cause damage to the apparatus, or restrict the air flow of the apparatus, or obstruct the normal operation of the apparatus.

Approved self-contained compressed air breathing apparatus may be used with approved cylinders from other approved self-contained compressed air breathing apparatus provided that such cylinders are of the same capacity and pressure rating. All compressed air cylinders used with self-contained breathing apparatus shall meet DOT and NIOSH criteria.

Self-contained breathing apparatuses must have a minimum service-life rating of 30 minutes in accordance with the methods and requirements specified by NIOSH under 42 CFR part 84, except for escape self-contained breathing apparatus (ESCBAs) used only for emergency escape purposes.

Self-contained breathing apparatus shall be provided with an indicator which automatically sounds an audible alarm when the remaining service life of the apparatus is reduced to within a range of 20 to 25 percent of its rated service time.

Positive-pressure breathing apparatus.

The following regulations for breathing apparatus relate more to design and specification than to maintenance. However, in order for the respirators to be in compliance with previously described regulations, they will need to be inspected regularly and those inspections will need to be recorded.

The employer shall assure that self-contained breathing apparatus ordered or purchased after July 1, 1981, for use by fire brigade members performing interior structural fire fighting operations, are of the pressure-demand or other positive-pressure type. Effective July 1, 1983, only pressure-demand or other positive-pressure self-contained breathing apparatus shall be worn by fire brigade members perform ing interior structural fire fighting.

This paragraph does not prohibit the use of a self-contained breathing apparatus where the apparatus can be switched from a demand to a positive-pressure mode. However, such apparatus shall be in the positive-pressure mode when fire brigade members are performing interior structural fire fighting operations.

Fire extinguishers.

This next section looks at provisions that regulate the portable fire extinguishers in a plant.

Selection and distribution.

Portable fire extinguishers shall be provided for employee use and selected and distributed based on the classes of anticipated workplace fires and on the size and degree of hazard which would affect their use.

The employer shall distribute portable fire extinguishers for use by employees on Class A fires so that the travel distance for employees to any extinguisher is 75 feet (22.9 m) or less.

The employer shall distribute portable fire extinguishers for use by employees on Class B fires so that the travel distance from the Class B hazard area to any extinguisher is 50ft. (15.2 m) or less.

The employer shall distribute portable fire extinguishers used for Class C hazards on the basis of the appropriate pattern for the existing Class A or Class B hazards.

The employer shall distribute portable fire extinguishers or other containers of Class D extinguishing agent for use by employees so that the travel distance from the combustible metal working area to any extinguishing agent is 75 feet (22.9 m) or less. Portable fire extinguishers for Class D hazards are required in those combustible metal working areas where combustible metal powders, flakes, shavings, or similarly sized products are generated at least once every two weeks.

Inspection, maintenance and testing.

The employer shall be responsible for the inspection, maintenance and testing of all portable fire extinguishers in the workplace.

> Portable extinguishers or hose used in lieu thereof under paragraph (d)(3) of this section shall be visually inspected monthly.
>
> The employer shall assure that portable fire extinguishers are subjected to an annual maintenance check. Stored pressure extinguishers do not require an internal examination. The employer shall record the annual maintenance date and retain this record for one year after the last entry or the life of the shell, whichever is less. The record shall be available to the Assistant Secretary upon request.
>
> The employer shall assure that stored pressure dry chemical extinguishers that require a 12-year hydrostatic test are emptied and subjected to applicable maintenance procedures every 6 years. Dry chemical extinguishers having non-refillable disposable containers are exempt from this requirement. When recharging or hydrostatic testing is performed, the 6-year requirement begins from that date.
>
> The employer shall assure that alternate equivalent protection is provided when portable fire extinguishers are removed from service for maintenance and recharging.

The regulations specify specific distances for placement of the extinguishers. They also specify the inspection and record keeping requirements. In order to maintain compliance, the company must insure that these records are kept and are available during an inspection. In some organizations, a separate safety department cares for these requirements. In other companies, this responsibility rests with the maintenance department.

Training and education.

As before, training requirements are highlighted in this section. Tracking and documenting the training is important for maintaining compliance with the regulations.

> Where the employer has provided portable fire extinguishers for employee use in the workplace, the employer shall also provide an educational program to familiarize employees with the general principles of fire extinguisher use and the hazards involved with incipient stage fire fighting.

The employer shall provide the education required in this section upon initial employment and at least annually thereafter.

The employer shall provide employees who have been designated to use fire fighting equipment as part of an emergency action plan with training in the use of the appropriate equipment.

The employer shall provide the training required in this section upon initial assignment to the designated group of employees and at least annually thereafter.

Maintenance.

The employer shall assure that water supply tanks are kept filled to the proper level except during repairs. When pressure tanks are used, the employer shall assure that proper pressure is maintained at all times except during repairs.

The employer shall assure that valves in the main piping connections to the automatic sources of water supply are kept fully open at all times except during repair.

The employer shall assure that hose systems are inspected at least annually and after each use to assure that all of the equipment and hose are in place, available for use, and in serviceable condition.

When the system or any portion thereof is found not to be serviceable, the employer shall remove it from service immediately and replace it with equivalent protection such as extinguishers and fire watches.

The employer shall assure that hemp or linen hose on existing systems is unracked, physically inspected for deterioration, and reracked using a different fold pattern at least annually. The employer shall assure that defective hose is replaced.

The employer shall designate trained persons to conduct all inspections required under this section.

These requirements for the fire system are generally a part of the preventive maintenance program. However, for specialty issues, like the reracking of the fire hose, specialty training may be required.

Therefore, if the hose inspection is part of the PM program, a skilled individual would have to be assigned to the inspection.

Sprinkler Systems

In all facilities and in most plants, sprinkler systems are used to reduce fire hazards. The proper maintenance of these systems is essential to insure proper operation when they are required to be utilized. This is critical, since systems not utilized at regular intervals may become unreliable. The inspections and service detailed in the following section are to insure reliability.

Maintenance.

> The employer shall properly maintain an automatic sprinkler system installed to comply with this section. The employer shall assure that a main drain flow test is performed on each system annually. The inspector's test valve shall be opened at least every two years to assure that the sprinkler system operates properly.

Sprinkler alarms.

> On all sprinkler systems having more than twenty (20) sprinklers, the employer shall assure that a local waterflow alarm is provided which sounds an audible signal on the premises upon water flow through the system equal to the flow from a single sprinkler.

Sprinkler spacing.

> The employer shall assure that sprinklers are spaced to provide a maximum protection area per sprinkler, a minimum of interference to the discharge pattern by building or structural members or building contents and suitable sensitivity to possible fire hazards. The minimum vertical clearance between sprinklers and material below shall be 18 inches (45.7 cm).

Chemical Fire Extinguisher Systems

In addition to sprinkler systems, some companies will utilize fixed chemical fire extinguishing systems. These systems are used for specific area coverage, similar to sprinkler systems. However, due to the fact that they eliminate oxygen from the area, they have their own set of regulations that must be considered. To insure proper operation, maintenance and inspection requirements are detailed.

> The employer shall assure that inspection and maintenance dates are recorded on the container, on a tag attached to the container, or in a central location. A record of the last semi-annual check shall be maintained until the container is checked again or for the life of the container, whichever is less.
>
> The employer shall train employees designated to inspect, maintain, operate, or repair fixed extinguishing systems and annually review their training to keep them up-to-date in the functions they are to perform.
>
> The employer shall sample the dry chemical supply of all but stored pressure systems at least annually to assure that the dry chemical supply is free of moisture which may cause the supply to cake or form lumps.
>
> The employer shall assure that the rate of application of dry chemicals is such that the designed concentration of the system will be reached within 30 seconds of initial discharge.
>
> The employer shall provide a distinctive pre-discharge employee alarm capable of being perceived above ambient light or noise levels when agent design concentrations exceed the maximum safe level for employee exposure. A pre-discharge employee alarm for alerting employees before system discharge shall be provided on Halon 1211 and carbon dioxide systems with a design concentration of 4 percent or greater and for Halon 1301 systems with a design concentration of 10 percent or greater. The pre-discharge employee alarm shall provide employees time to safely exit the discharge area prior to system discharge
>
> Where egress from an area cannot be accomplished within one minute, the employer shall not use Halon 1301 in concentrations greater than 7 percent
>
> Where egress takes greater than 30 seconds but less than one

minute, the employer shall not use Halon 1301 in a concentration greater than 10 percent.

Halon 1301 concentrations greater than 10 percent are only permitted in areas not normally occupied by employees provided that any employee in the area can escape within 30 seconds. The employer shall assure that no unprotected employees enter the area during agent discharge.

The employer shall restore all fire detection systems and components to normal operating condition as promptly as possible after each test or alarm. Spare detection devices and components which are normally destroyed in the process of detecting fires shall be available on the premises or from a local supplier in sufficient quantities and locations for prompt restoration of the system.

As previous noted, fire extinguishing systems need inspections and service at specified intervals. If they are covered by the PM program, only trained and certified individuals should perform the inspections.

Fire Detection Systems

Plants and facilities are required to have fire detection and prevention systems installed. While proper design and installation is essential to insure protection for the plant or facility, proper maintenance is essential to insure the systems function correctly if there ever is a fire. The following regulations detail these requirements.

Maintenance and testing.

The employer shall maintain all systems in an operable condition except during repairs or maintenance.

The employer shall assure that fire detectors and fire detection systems are tested and adjusted as often as needed to maintain proper reliability and operating condition except that factory calibrated detectors need not be adjusted after installation.

The employer shall assure that pneumatic and hydraulic operated detection systems installed after January 1, 1981, are equipped with supervised systems.

The employer shall assure that the servicing, maintenance and testing of fire detection systems, including cleaning and necessary sensitivity adjustments are performed by a trained person knowledgeable in the operations and functions of the system.

The employer shall also assure that fire detectors that need to be cleaned of dirt, dust, or other particulates in order to be fully operational are cleaned at regular periodic intervals.

Protection of fire detectors.

The employer shall assure that fire detection equipment installed outdoors or in the presence of corrosive atmospheres be protected from corrosion. The employer shall pro-vide a canopy, hood, or other suitable protection for detection equipment requiring protection from the weather.

The employer shall locate or otherwise protect detection equipment so that it is protected from mechanical or physical impact which might render it inoperable.

The employer shall assure that detectors are supported independently of their attachment to wires or tubing.

Response time.

The employer shall assure that fire detection systems installed for the purpose of actuating fire extinguishment or suppression systems shall be designed to operate in time to control or extinguish a fire.

The employer shall assure that fire detection systems installed for the purpose of employee alarm and evacuation be designed and installed to provide a warning for emergency action and safe escape of employees.

The employer shall not delay alarms or devices initiated by fire detector actuation for more than 30 seconds unless such delay is necessary for the immediate safety of employees. When such delay is necessary, it shall be addressed in an emergency action plan meeting the requirements of Number, location and spacing of detecting devices. The employer shall assure that the number, spacing and location of fire detectors is based upon design data obtained from field experience, or tests, engineering surveys, the manufacturer's recommendations, or a recognized testing laboratory listing.

Emergency Alarms

Emergency alarms are devices used to warn employees before an automatic fire suppression system is activated. This alarm and its interval are designed to give the employees time to vacate the area before the system is discharged. It is critical that these alarms function correctly to prevent employees from being trapped in the area.

Scope and application.

This section applies to all emergency employee alarms installed to meet a particular OSHA standard. This section does not apply to those discharge or supervisory alarms required on various fixed extinguishing systems or to supervisory alarms on fire suppression, alarm or detection systems unless they are intended to be employee alarm systems.

The requirements in this section that pertain to maintenance, testing and inspection shall apply to all local fire alarm signaling systems used for alerting employees regardless of the other functions of the system.

General requirements.

The employee alarm system shall provide warning for necessary emergency action as called for in the emergency action plan, or for reaction time for safe escape of employees from the workplace or the immediate work area, or both.

The employee alarm shall be capable of being perceived above ambient noise or light levels by all employees in the affected portions of the workplace. Tactile devices may be used to alert those employees who would not otherwise be able to recognize the audible or visual alarm

The employee alarm shall be distinctive and recognizable as a signal to evacuate the work area or to perform actions designated under the emergency action plan.

The employer shall explain to each employee the preferred means of reporting emergencies, such as manual pull box alarms, public address systems, radio or telephones. The employer shall post emergency telephone numbers near telephones, or employee notice boards, and other conspicuous locations when telephones serve as a means of reporting emergencies. Where a communication system also serves as the employee alarm system, all emergency messages shall have priority over all non-emergency messages.

The employer shall establish procedures for sounding emergency alarms in the workplace. For those employers with 10 or fewer employees in a particular workplace, direct voice communication is an acceptable procedure for sounding the alarm provided all employees can hear the alarm. Such workplaces need not have a back-up system.

Installation and restoration.

The employer shall assure that all devices, components, combinations of devices or systems constructed and installed to comply with this standard are approved. Steam whistles, air horns, strobe lights or similar lighting devices, or tactile devices meeting the requirements of this section are considered to meet this requirement for approval.

The employer shall assure that all employee alarm systems are restored to normal operating condition as promptly as possible after each test or alarm. Spare alarm devices and components subject to wear or destruction shall be available in sufficient quantities and locations for prompt restoration of the system.

Maintenance and testing.

The employer shall assure that all employee alarm systems are maintained in operating condition except when undergoing repairs or maintenance.

The employer shall assure that a test of the reliability and adequacy of non-supervised employee alarm systems is made every two months. A different actuation device shall be used in each test of a multi-actuation device system so that no individual device is used for two consecutive tests.

The employer shall maintain or replace power supplies as often as is necessary to assure a fully operational condition. Back-up means of alarm, such as employee runners or telephones, shall be provided when systems are out of service.

The employer shall assure that employee alarm circuitry installed after January 1, 1981, which is capable of being supervised is supervised and that it will provide positive notification to assigned personnel whenever a deficiency exists in the system. The employer shall assure that allsupervised employee alarm systems are tested at least annually for reliability and adequacy.

The employer shall assure that the servicing, maintenance and testing of employee alarms are done by persons trained in the designed operation and functions necessary for reliable and safe operation of the system.

Manual operation.

The employer shall assure that manually operated actuation devices for use in conjunction with employee alarms are unobstructed, conspicuous and readily accessible.

For plants where any of the fire systems are the responsibility of the maintenance department, the required personnel must be adequately trained and monitored. It is possible that in some plants these responsibilities are shared by the maintenance and operations groups. In this instance, the duties must be clearly defined. There can be no ambiguity. In addition, the responsibilities for record keeping must also be clearly understood.

Industrial Trucks

Industrial trucks are equipment typically designated for material handling duties in most plants. While there are other uses, such as for movement of maintenance parts, the majority of the industrial trucks are operated by production personnel.

Industrial Truck Operator Training

Since most production operators do not have a technical background, it is necessary to provide them with training to insure they can operate the industrial truck safely. In addition, there are certain routine tasks that need to be performed by the operator. It is necessary to provide the training to insure these tasks are performed correctly and safely.

Changing and charging storage batteries.

Battery charging installations shall be located in areas designated for that purpose.

Facilities shall be provided for flushing and neutralizing spilled electrolyte, for fire protection, for protecting charging apparatus from damage by trucks, and for adequate ventilation for dispersal of fumes from gassing batteries.

A conveyor, overhead hoist, or equivalent material handling equipment shall be provided for handling batteries.

OSHA requires that "reinstalled batteries shall be properly positioned and secured in the truck [emphasis added]." Accordingly, batteries in all electric powered industrial trucks covered by the standard must be secured in place both horizontally and vertically when the truck is in use.

Care shall be taken to assure that vent caps are function ing. The battery (or compartment) cover(s) shall be open to dissipate heat.

Tools and other metallic objects shall be kept away from the top of uncovered batteries.

Inspections.

OSHA does have a general requirement that all powered industrial trucks be examined before being placed in service. This examination is required daily or after each shift if the trucks are used on a round-the-clock basis.

The standard does not require documentation of a powered industrial truck examination. Therefore, it would be at the employer's discretion to determine the duration of powered industrial truck examination record retention. The employer must not place a powered industrial truck in service if the examination shows any condition which adversely affects the safety of the vehicle. Defects found during an examination must be immediately reported and corrected.

Operator Qualifications

Employers employing new operators or temporary labor operators who claim prior training would be required to evaluate the applicability and adequacy of prior training to determine if all required training topics have been covered.

The employer may wish to consider these factors:

the type of equipment the operator has operated;

how much experience the operator has had on that equipment;

how recently this experience was gained; and

the type of environment in which the operator worked.

The employer may, but is not required to, use written documentation of the earlier training to determine whether an operator has been properly trained.

The operator's competency may also simply be evaluated by the employer or another person with the requisite knowledge, skills, and experience to perform evaluations. The employer can determine from this information whether the experience is recent and thorough enough, and whether the operator has demonstrated sufficient comptence in operating in the powered industrial truck to forego any or some of the initial training. Some training on site regarding specificfactors of the newoperator's workplace is likely always to be necessary.

OSHA requires that employers certify operator that each has been trained and evaluated as required by paragraph (l). Certification includes the name of an operator, the date of the training, the date of the evaluation, and the identity of the person(s) performing the training or evaluation.

Operator certification.

Employers are required to certify that each operator has been trained and evaluated. As required by OSHA, the certification shall include the name of the operator, the date of the training, the date of the evaluation, and the identity of the person(s) performing the training or evaluation.

Maintaining Records.

OSHA does not specify who must maintain the records; however, the employer is ultimately responsible for ensuring the availability of these records. Third-party trainers who agree to maintain the records would also need to ensure their immediate availability to the employer.

Therefore, an operator must be trained and evaluated in the safe operation for the type of truck that the operator will be assigned to operate in the employer's workplace. For example, if an operator is assigned to operate a sit-down counterbalanced rider truck, then the operator must be trained and evaluated in the safe operation for that type of truck. If an operator is assigned to operate an operator-up counterbalanced front/side loader truck, or a rough terrain forklift, then the operator must be trained and evaluated in the safe operation for those types of trucks.

A sit-down counterbalanced rider truck, an operator-up counter balanced front/side loader truck, and a rough terrain forklift are different types of trucks. Operators who have successfully completed training and evaluation (in a specific type of truck) would not need additional training when they are assigned to operate the same type of truck made by a different manufacturer. However, operators would need additional training if the applicable truck-related and workplace related topics are different for that truck.

Operator training

To this point, the discussion has centered around what topics should be covered in the training. However, the regulations also specify some of the conditions under which the training should be conducted, including actual content and method of delivery. The following regulations highlight these points.

Safe operation.

OSHA's enforcement policy relative to the use of seat belts on powered industrial trucks is that employers are obligated to require operators of powered industrial trucks which are equipped with operator restraint devices or seat belts to use the devices.

The employer shall ensure that each powered industrial truck operator is competent to operate a powered industrial truck safely, as demonstrated by the successful completion of the training and evaluation.

Prior to permitting an employee to operate a powered industrial truck (except for training purposes), the employer shall ensure that each operator has successfully completed the training required by this section.

Training program implementation.

Trainees may operate a powered industrial truck only:

Under the direct supervision of persons who have the knowledge, training, and experience to train operators and evaluate their competence; and

Where such operation does not endanger the trainee or other employees.

Training shall consist of a combination of formal instruction (e.g., lecture, discussion, interactive computer learning, videotape, written material), practical training (demonstrations performed by the trainer and practical exercises performed by the trainee), and evaluation of the operator's performance in the workplace.

All operator training and evaluation shall be conducted by persons who have the knowledge, training, and experience to train powered industrial truck operators and evaluate their competence.

Training program content.

Powered industrial truck operators shall receive initial training in the following topics, except in topics which the employer can demonstrate are not applicable to safe operation of the truck in the employer's workplace.

Truck-related topics:

Operating instructions, warnings, and precautions for the types of truck the operator will be authorized to operate;

Differences between the truck and the automobile;

Truck controls and instrumentation: where they are located, what they do, and how they work;

Engine or motor operation;

Steering and maneuvering;

Visibility (including restrictions due to loading);

Fork and attachment adaptation, operation, and use limitations;

Vehicle capacity;

Vehicle stability;

Any vehicle inspection and maintenance that the operator will be required to perform;

Refueling and/or charging and recharging of batteries;

Operating limitations;

Any other operating instructions, warnings, or precautions listed in the operator's manual for the types of that vehicle the employee is being trained to operate.

Workplace-related topics:

Surface conditions where the vehicle will be operated;

Composition of loads to be carried and load stability;

Load manipulation, stacking, and unstacking;

Pedestrian traffic in areas where the vehicle will be operated;

Narrow aisles and other restricted places where the vehicle will be operated;

Hazardous (classified) locations where the vehicle will be operated;

Ramps and other sloped surfaces that could affect the vehicle's stability;

Closed environments and other areas where insufficient ventilation or poor vehicle maintenance could cause a buildup of carbon monoxide or diesel exhaust;

Other unique or potentially hazardous environmental conditions in the workplace that could affect safe operation.

Refresher training and evaluation.

Refresher training, including an evaluation of the effectiveness of that training, shall be conducted as required by OSHA to ensure that the operator has the knowledge and skills needed to operate the powered industrial truck safely.

Refresher training in relevant topics shall be provided to the operator when:

The operator has been observed to operate the vehicle in an unsafe manner;

The operator has been involved in an accident or near-miss incident;

The operator has received an evaluation that reveals that the operator is not operating the truck safely;

The operator is assigned to drive a different type of truck; or A condition in the workplace changes in a manner that could affect safe operation of the truck.

An evaluation of each powered industrial truck operator's performance shall be conducted at least once every three years.

Certification.

The employer shall certify that each operator has been trained and evaluated as required by this paragraph. The certification shall include the name of the operator, the date of the training, the date of the evaluation, and the identity of the person(s) performing the training or evaluation.

Unattended Trucks.

A powered industrial truck is unattended when the operator is 25 ft. or more away from the vehicle which remains in his view, or whenever the operator leaves the vehicle and it is not in his view.

When the operator of an industrial truck is dismounted and within 25 ft. of the truck still in his view, the load engaging means shall be fully lowered, controls neutralized, and the brakes set to prevent movement.

Lifting Personnel.

Whenever a truck is equipped with vertical only, or vertical and horizontal controls elevatable with the lifting carriage or forks for lifting personnel, the following additional precautions shall be taken for the protection of personnel being elevated. Use of a safety platform firmly secured to the lifting carriage and/or forks.

Means shall be provided whereby personnel on the platform can shut off power to the truck.
Such protection from falling objects as indicated necessary by the operating conditions shall be provided.

Maintenance of industrial trucks.

In addition to the requirements for operations, there are specific maintenance issues that are impacted by the regulations. These are typically safety related for service on the industrial truck, but may also impact where the service may take place.

> Any power-operated industrial truck not in safe operating condition shall be removed from service. All repairs shall be made by authorized personnel.
>
> Those repairs to the fuel and ignition systems of industrial trucks which involve fire hazards shall be conducted only in locations designated for such repairs.

Trucks in need of repairs to the electrical system shall have the battery disconnected prior to such repairs.

All parts of any such industrial truck requiring replacement shall be replaced only by parts equivalent as to safety with those used in the original design.

Industrial trucks shall not be altered so that the relative positions of the various parts are different from what they were when originally received from the manufacturer, nor shall they be altered either by the addition of extra parts not provided by the manufacturer or by the elimination of any parts. Additional counterweighting of fork trucks shall not be done unless approved by the truck manufacturer.

Industrial trucks shall be examined before being placed in service, and shall not be placed in service if the examination shows any condition adversely affecting the safety of the vehicle. Such examination shall be made at least daily. Where industrial trucks are used on a round-the-clock basis, they shall be examined after each shift. Defects when found shall be immediately reported and corrected.

Water mufflers shall be filled daily or as frequently as is necessary to prevent depletion of the supply of water below 75 percent of the filled capacity. Vehicles with mufflers having screens or other parts that may become clogged shall not be operated while such screens or parts are clogged. Any vehicle that emits hazardous sparks or flames from the exhaust system shall immediately be removed from service, and not returned to service until the cause for the emission of such sparks and flames has been eliminated.

When the temperature of any part of any truck is found to be in excess of its normal operating temperature, thus creating a hazardous condition, the vehicle shall be removed from service and not returned to service until the cause for such overheating has been eliminated.

Industrial trucks shall be kept in a clean condition, free of lint, excess oil, and grease. Noncombustible agents should be used for cleaning trucks. Low flash point (below 100 deg. F.) solvents shall not be used. High flash point (at or above 100 deg. F.) solvents may be used. Precautions regarding toxicity, ventilation, and fire hazard shall be consonant with the agent or solvent used.

Specific Equipment

OSHA regulations in particular specify certain operational and maintenance guidelines for specific industries and types of equipment. This section highlights several of these guidelines. The purpose of this section is to highlight that all equipment is impacted by the OSHA regulations and that many of the regulations apply in one or more industries.

BAKERY EQUIPMENT

Bakery equipment is food processing related, but still has many of the basic components of other industrial systems. The following are some regulations specific to bakery equipment, but can cross over into other industries. So while this may not apply directly to petrochemical industries (for example), these regulations can be cited as examples of what should be done to protect employees in a petrochemical plant.

Sprockets and V-belt drives.

Sprockets and V-belt drives located within reach from platforms or passageways or located within 8 feet 6 inches from the floor shall be completely enclosed

Lubrication.

Where machinery must be lubricated while in motion, stationary lubrication fittings inside a machine shall be provided with extension piping to a point of safety so that the employee will not have to reach into any dangerous part of the machine when lubricating.

Hot pipes.

Exposed hot water and steam pipes shall be covered with insulating material wherever necessary to protect employee from contact.

Man lifts.

Man lifts shall be prohibited in bakeries. Bag or barrel lifts shall not be used as man lifts.

Chain tackle.

All chain tackle shall be marked prominently, permanently, and legibly with maximum load capacity.

All chain tackle shall be marked permanently and legibly with minimum support specification.

Sugar and spice pulverizers.

All drive belts used in connection with sugar and spice pulverizers shall be grounded by means of metal combs or other effective means of removing static electricity. All pulverizing of sugar or spice grinding shall be done in accordance with NFPA 62-1967 (Standard for Dust Hazards of Sugar and Cocoa) and NFPA 656-1959 (Standard for Dust Hazards in Spice Grinding Plants), which are incorporated by reference as specified in Sec. 1910.6.

Bakery Ovens

Ovens shall be located so that possible fire or explosion will not expose groups of persons to possible injury. For this reason ovens shall not adjoin lockers, lunch or sales rooms, main passageways, or exits.

All safety devices on ovens shall be inspected at intervals of not less than twice a month by an especially appointed, properly instructed bakery employee, and not less than once a year by representatives of the oven manufacturers.

Duct systems (in ovens) operating under pressure shall be tested for tightness in the initial starting of the oven and also at intervals not farther apart than 6 months.

Steam Line Insulation

The following question and answer regarding temperatures of metal pipes is taken from an interpretation of standards developed for the bakery industry. Note that this industry specific standard (1910.261(b)(5), and others like it, overrides the more generic OSHA standard.

"At what temperature should metal pipe be insulated to avoid burning of the skin on contact?"

Steam and hot-water pipes.

All exposed steam and hot-water pipes within 7 feet of the floor or working platform or within 15 inches measured horizontally from stairways, ramps, or fixed ladders shall be covered with an insulating material, or guarded in such manner as to prevent contact.

Steam pipes.

All pipes carrying steam or hot water for process or servicing machinery, when exposed to contact and located with seven feet of the floor or working platform shall be covered with a heat-insulating material, or otherwise properly guarded."

"Regardless of height, open-sided floors, walkways, platforms, or runaways above or adjacent to dangerous equipment, pickling or galvanizing tanks, degreasing units, and similar hazards shall be guarded with a standard railing and toe board."

The control of hazardous energy (lockout/tagout) standard, covers hazardous energy, including thermal, during the servicing and maintenance of machines or equipment. Thermal energy may be dissipated or controlled and it is the result of mechanical work, radiation, or electrical resistance. This standard addresses practices and procedures that are necessary to disable machinery or equipment and to prevent the release of potentially hazardous energy while maintenance and servicing activities are performed.

"Protective equipment, including personal protective equipment for eyes, face, head, and extremities, protective clothing, respiratory devices, and protective shields and barriers, shall be provided, used, and maintained in a sanitary and reliable condition wherever it is necessary by reason of hazards of processes or environment, chemical hazards, radiological hazards, or mechanical irritants encountered in a manner capable of causing injury or impairment in the function of any part of the body through absorption, inhalation or physical contact."

The personal protective equipment standard would apply to hot surfaces where the hazards have not been eliminated through engineering or administrative controls. This standard requires employers to assess the workplace to determine if hazards that require the use of PPE are present or are likely to be present. The employer must select and have affected employees use properly fitted PPE suitable for protection against these hazards, as well as provide employee training and conduct periodic inspections to assure procedures are being followed. Suitable thermal protection would be necessary to provide employees with thermal insulation from hazardous hot pipe surfaces.

Each employer shall furnish to each of his employees employment and a place of employment which are free from recognized hazards that are causing or are likely to cause death or serious physical harm to his employees.

Sawmills

Equipment used in sawmills is also found in several other industries, particularly in pulp and paper operations. As before, many of these regulations are found only in the OSHA section on sawmills, but will have application in other industries.

Floor maintenance.

The flooring of buildings, docks, and passageways shall be kept in good repair. When a hazardous condition develops that cannot be immediately repaired, the area shall be guarded until adequate repairs are made.

Nonslip floors.

Floors, footwalks, and passageways in the work area around machines or other places where a person is required to stand or walk shall be provided with effective means to minimize slipping.

Stairways

Construction.

Stairways shall be constructed in accordance with 1910.24.

Handrails.

Stairways shall be provided with a standard handrail on at least one side or on any open side. Where stairs are more than four feet wide there shall be a standard handrail at each side, and where more than eight feet wide, a third standard handrail shall be erected in the center of the stairway

Swinging doors.

All swinging doors shall be provided with windows; with one window for each section of double swinging doors. Such windows shall be of shatterproof or safety glass unless otherwise protected against breakage.

Hydraulic systems in Sawmills.

Maintenance requirements for hydraulic systems in sawmills focus on protecting employees from system activation or drift during maintenance. The following regulation is an example.

Means shall be provided to block, chain, or otherwise secure equipment normally supported by hydraulic pressure so as to provide for safe maintenance.

Ropes or cables

Since there are many heavy objects in a sawmill, the use of wire rope is common. The following are some regulations for wire rope and related equipment written specifically for the sawmills. The end of the section also contains some specifics about chain usage.

Wire rope or cable shall be inspected when installed and once each week thereafter, when in use. It shall be removed from hoisting or load-carrying service when kinked or when one of the following conditions exists:

When three broken wires are found in one lay of 6 by 6 wire rope.

When six broken wires are found in one lay of 6 by 19 wire rope.

When nine broken wires are found in one lay of 6 by 37 wire rope.

When eight broken wires are found in one lay of 8 by 19 wire rope.

When marked corrosion appears.

Wire rope of a type not described herein shall be remove from service when 4 percent of the total number of wires composing such rope are found to be broken in one lay.

Wire rope removed from service due to defects shall be plainly marked or identified as being unfit for further use on cranes, hoists, and other load-carrying devices.

The ratio between the rope diameter and the drum, block, sheave, or pulley tread diameter shall be such that the rope will adjust itself to the bend without excessive wear, deformation, or injury. In no case shall the safe value of drums, blocks, sheaves, or pulleys be reduced when replacing such items unless compensating changes are made for rope used and for safe loading limits.

Drums, sheaves, and pulleys.

Drums, sheaves, and pulleys shall be smooth and free from surface defects liable to injure rope. Drums, sheaves, or pulleys having eccentric bores or cracked hubs, spokes, or flanges shall be removed from service.

Connections.

Connections, fittings, fastenings, and other parts used in connection with ropes and cables shall be of good quality and of proper size and strength, and shall be installed in accordance with the manufacturer's recommendations.

Socketing, splicing, and seizing.

Socketing, splicing, and seizing of cables shall be performed only by qualified persons.

All eye splices shall be made in an approved manner and wire rope thimbles of proper size shall be fitted in the eye, except that in slings the use of thimbles shall be optional.

Wire rope clips attached with U-bolts shall have these bolts on the dead or short end of the rope. The U-bolt nuts shall be tightened immediately after initial load carrying use and at frequent intervals thereafter.

When a wedge socket-type fastening is used, the dead or short end of the cable shall be clipped with a U-bolt or otherwise made secure against loosening.

Fittings.

Hooks, shackles, rings, pad eyes, and other fittings that show excessive wear or that have been bent, twisted, or otherwise damaged shall be removed from service.

Running lines.

Running lines of hoisting equipment located within 6 feet 6 inches of the ground or working level shall be boxed off or otherwise guarded, or the operating area shall be restricted.

Number of wraps on drum.

There shall be not less than two full wraps of hoisting cable on the drum of cranes and hoists at all times of operation.

Drum flanges.

Drums shall have a flange at each end to prevent the cable from slipping off.

Sheave guards.

Bottom sheaves shall be protected by close fitting guards to prevent cable from jumping the sheave.

Preventing abrasion.

The reeving of a rope shall be so arranged as to minimize chafing or abrading while in use.

Chains

Chains used in load carrying service shall be inspected before initial use and weekly thereafter.

Chain shall be normalized or annealed periodically as recommended by the manufacturer.

If at any time any 3-foot length of chain is found to have stretched one-third the length of a link it shall be discarded.

Bolts or nails shall not be placed between two links to shorten or join chains.

Broken chains shall not be spliced by inserting a bolt between two links with the head of the bolt and nut sustaining the load, or by passing one link through another and inserting a bolt or nail to hold it.

Miscellaneous Sawmill Equipment

The following items that are regulated. Again, while these specific to sawmills, other industries use stackers and personal transportation vehicles. These regulations can apply to these industries also.

Inspection.

Every stacker and unstacker shall be inspected at frquent intervals and all defective parts shall be immediately repaired or replaced

Operation in buildings.

Vehicles powered by internal combustion engines shall not operate in buildings unless the buildings are adequately ventilated.

Chapter 6

LOGGING OPERATIONS

Logging operations are also a narrowly defined vertical industry segment. However, there are some interesting regulations impacting hand and power tool usage and general training requirements for employees and supervisors. These are highlighted in the next section.

Hand and portable powered tools

While hand and power tools have been considered previously, these regulations are specific to the logging operations. The cross application of the regulations are quite interest ing. Consider the tool inspections and service in the next section.

General requirements.

The employer shall assure that each hand and portable powered tool, including any tool provided by an employee, is maintained in serviceable condition.

The employer shall assure that each tool, including any tool provided by an employee, is inspected before initial use during each workshift. At a minimum, the inspection shall include the following:

Handles and guards, to assure that they are sound, tight-fitting, properly shaped, free of splinters and sharp edges, and in place;

The employer shall assure that operating and maintenance instructions are available on the machine or in the area where the machine is being operated. Each machine operator and maintenance employee shall comply with the operating and maintenance instructions.

No employee other than the operator shall ride on any mobile machine unless seating, seat belts and other protection equivalent to that provided for the operator are provided

Transparent material shall be kept clean to assure operator visibility.

Transparent material that may create a hazard for the operator, such as but not limited to, cracked, broken or scratched safety glass, shall be replaced.

Training

In addition to tool inspection and service, there are also specific training requirements for tools and their safe usage. The training requirements are so detailed, they even discuss the structure of the training. This is shown in the following section.

The employer shall provide training for each employee, including supervisors, at no cost to the employee.

Frequency.

Training shall be provided as follows:

As soon as possible but not later than the effective date of this section for initial training for each current and new employee;

Prior to initial assignment for each new employee;

Whenever the employee is assigned new work tasks, tools, equipment, machines or vehicles; and

Whenever an employee demonstrates unsafe job perfomance.

Content.

At a minimum, training shall consist of the following elements:

Safe performance of assigned work tasks;

Safe use, operation and maintenance of tools, machines and vehicles the employee uses or operates, including emphasis on understanding and following the manufacturer's operating and maintenance instructions, warnings and precautions;

Recognition of safety and health hazards associated with the employee's specific work tasks, including the use of measures and work practices to prevent or control those hazards;

Recognition, prevention and control of other safety and health hazards in the logging industry;

Procedures, practices and requirements of the employer's work site; and

The requirements of this standard.
Training of an employee due to unsafe job performance, or assignment of new work tasks, tools, equipment, machines, or vehicles; may be limited to those elements of this section which are relevant to the circumstances giving rise to the need for training.

Portability of training.
Each current employee who has received training in the particular elements specified in paragraph (i)(3) of this section shall not be required to be retrained in those elements.
Each new employee who has received training in the particular elements specified in this section shall not be required to be retrained in those elements prior to initial assignment.
The employer shall train each current and new employee in those elements for which the employee has not received training.
The employer is responsible for ensuring that each current and new employee can properly and safely perform the work tasks and operate the tools, equipment, machines, and vehicles used in their job.
Each new employee and each employee who is required to be trained as specified in paragraph of this section, shall work under the close supervision of a designated person until the employee demonstrates to the employer the ability to safely perform their new duties independently.

The following requirement has existed since the standard 29 CFR 1910.263 from OSHA was adopted on May 29, 1971.

TELECOMMUNICATIONS AND UTILITIIES

One final OSHA regulation from the telecommunication industry deal with rotating equipment guarding. While some guidelines are provided in the generic section of the standards, this one is specific to the power plant machinery.

> Equipment, machinery and machine guarding. When power plant machinery in telecommunications centers is operated with commutators and couplings uncovered, the adjacent housing shall be clearly marked to alert personnel to the rotating machinery.

FDA REGULATED PLANTS

The FDA has specific regulations dealing with personnel and their work practices in FDA-regulated plants. These are similar in scope to the OSHA guidelines, but go beyond them in certain areas, such as clothing, hairstyle, and jewelry requirements. These regulations are highlighted in the following material.

Practices Related to Human Food.

The following are personnel practices that are directed to plants that produce human food products. These have a direct impact on the operational personnel.

Personnel.

The plant management shall take all reasonable measures and precautions to ensure the following:

(a) *Disease control.* Any person who, by medical examination or supervisory observation, is shown to have, or appears to have, an illness, open lesion, including boils, sores, or infected wounds, or any other abnormal source of microbial contamination by which there is a reasonable possibility of food, food-contact surfaces, or food-packaging materials becoming contaminated, shall be excluded from any operations which may be expected to result in such contamination until the condition is corrected. Personnel shall be instructed to report such health conditions to their supervisors.

(b) *Cleanliness.* All persons working in direct contact with food, food-contact surfaces, and food-packaging materials shall conform to hygienic practices while on duty to the extent necessary to protect against contamination of food. The methods for maintaining cleanliness include, but are not limited to:

(1) Wearing outer garments suitable to the operation in a manner that protects against the contamination of food, food-contact surfaces, or food-packaging materials.

(2) Maintaining adequate personal cleanliness.

(3) Washing hands thoroughly (and sanitizing if necessary to protect against contamintion with undesirable microorganisms) in an adequate hand-washing facility before starting

work, after each absence from the work station, and at any other time when the hands may have become soiled or contaminated.

(4) Removing all unsecured jewelry and other objects that might fall into food, equipment, or containers, and removing hand jewelry that cannot be adequately sanitized during periods in which food is manipulated by hand. If such hand jewelry cannot be removed, it may be covered by material which can be maintained in an intact, clean, and sanitary condition and which effectively protects against the contamination by these objects of the food, food-contact surfaces, or food-packaging materials.

(5) Maintaining gloves, if they are used in food handling, in an intact, clean, and sanitary condition. The gloves should be of an impermeable material.

(6) Wearing, where appropriate, in an effective manner, hair nets, headbands, caps, beard covers, or other effective hair restraints.

(7) Storing clothing or other personal belongings in areas other than where food is exposed or where equipment or utensils are washed.

(8) Confining the following to areas other than where food may be exposed or where equipment or utensils are washed: eating food, chewing gum, drinking beverages, or using tobacco.

(9) Taking any other necessary precautions to protect against contamination of food, food-contact surfaces, or food-packaging materials with microorganisms or foreign substances including, but not limited to, perspiration, hair, cosmetics, tobacco, chemicals, and medicines applied to the skin.

(c) *Education and training.* Personnel responsible for identifying sanitation failures or food contamination should have a background of education or experience, or a combination thereof, to provide a level of competency necessary for production of clean and safe food. Food handlers and supervisors should receive appropriate training in proper food handling techniques and food-protection principles and should be informed of the danger of poor personal hygiene and insanitary practices.

(d) *Supervision.* Responsibility for assuring compliance by all personnel with all requirements of this part shall be clearly assigned to competent supervisory personnel.

Personnel qualifications.

(a) Each person engaged in the manufacture, processing, packing, or holding of a drug product shall have education, training, and experience, or any combination thereof, to enable that person to perform the assigned functions. Training shall be in the particular operations that the employee performs and in current good manufacturing practice (including the current good manufacturing practice regulations in this chapter and written procedures required by these regulations) as they relate to the employee's functions. Training in current good manufacturing practice shall be conducted by qualified individuals on a continuing basis and with sufficient frequency to assure that employees remain familiar with CGMP requirements applicable to them.

(b) Each person responsible for supervising the manufacture, processing, packing, or holding of a drug product shall have the education, training, and experience, or any combination thereof, to perform assigned functions in such a manner as to provide assurance that the drug product has the safety, identity, strength, quality, and purity that it purports or is represented to possess.

(c) There shall be an adequate number of qualified personnel to perform and supervise the manufacture, processing, packing, or holding of each drug product.

In all industries today, there is a management philosophy to involve operations personnel in maintenance activities. These activities range from simple visual inspections, to actual servicing of the equipment. Regardless of the type of activities the operations personnel are required to carry out on the equipment, regulatory agencies have established regulatory guidelines. So all managers should be aware that operators are impacted by the regulations and need to be trained and qualified, so as to avoid any regulatory violations.

CHAPTER 7

Reliability Centered Maintenance

True Reliability Centered Maintenance (RCM) deals with equipment life from concept to disposal. Reliability-centered maintenance is a systematic approach to developing a focused, effective, and cost efficient preventive and predictive maintenance program. The RCM technique is best initiated early in the equipment design process and should evolve as the equipment design, development, construction, commissioning, and operating activities progress. The technique, however, can also be used to evaluate preventive and predictive maintenance programs for existing equipment systems with the objective of continuously improving these systems.

The goals for an RCM program are as follows:

To achieve the maximum reliability, performance, and safety of the equipment.

Restore equipment to the required levels of performance when deterioration occurs (but before failure).

Collect the data (during the life of the equipment) to change the design of the equipment in order to improve its reliability.

Accomplish the above with minimal life-cycle costs.

The RCM technique was developed in the 1960s primarily through the efforts of the commercial airline industry. The essence of this technique is a structured decision tree, which leads the analyst through a tailored logic in order to outline the most applicable preventive maintenance tasks. There are two main applications for reliability centered maintenance: *equipment in the design phase and equipment already installed and being used.*

This chapter includes many regulations that should be considered during the entire life cycle of equipment and products.

Process Safety Management

Process safety management (PSM) is a series of regulations designed to insure the safety of the plant process equipment. Because of the hazards involved in producing certain products, PSM is one of the most important considerations for industry today. Annually there are several major incidents at plants in the United States that result in death or injury to employees that are related to mistakes made in areas impacted by PSM regulations. PSM regulations require strict attention to details by all employees who utilize a company's assets, including maintenance, operations, and engineering.

Process safety information

Process Safety Management regulations are a sub-set of the OSHA regulations. The documentation of all safety information and the hazards associated with a particular process is required to be detailed. Some examples are highlighted in the regulations shown in the following material.

> The employer shall complete a compilation of written process safety information before conducting any process hazard analysis required by the standard. The compilation of written process safety information is to enable the employer and the employees involved in operating the process to identify and understand the hazards posed by those processes involving highly hazardous chemicals. This process safety information shall include information pertaining to the hazards of the highly hazardous chemicals used or produced by the process, information pertaining to the technology of the process, and information pertaining to the equipment in the process.
>
> Information pertaining to the hazards of the highly hazard-

ous chemicals in the process. This information shall consist of at least the following:

Toxicity information;

Permissible exposure limits;

Physical data;

Reactivity data:

Corrosivity data;

Thermal and chemical stability data; and

Hazardous effects of inadvertent mixing of different materials that could foreseeably occur.

Note: Material Safety Data Sheets meeting the requirements of 29 CFR 1910.1200(g) may be used to comply with this requirement to the extent they contain the information required by this subparagraph.

Information Pertaining to the Technology of the Process.

Information concerning the technology of the process shall include at least the following:

A block flow diagram or simplified process flow diagram

Process chemistry;

Maximum intended inventory;

Safe upper and lower limits for such items as temperatures, pressures, flows or compositions; and,

An evaluation of the consequences of deviations, including those affecting the safety and health of employees.

Where the original technical information no longer exists, such information may be developed in conjunction with the process hazard analysis in sufficient detail to support the analysis.

Information Pertaining to the Equipment in the Process.

Information pertaining to the equipment in the process shall include:

Materials of construction;

Piping and instrument diagrams (P&ID's);

Electrical classification;

Relief system design and design basis;

Ventilation system design;

Design codes and standards employed;

Material and energy balances for processes built after May 26, 1992; and,

Safety systems (e.g. interlocks, detection or suppression systems).

The employer shall document that equipment complies with recognized and generally accepted good engineering practices.

As can be seen from the previous material, the PSM regulations require extensive documentation on the part of the maintenance organization. Prior to the equipment commissioning, it is necessary for the design engineering group to record all of the equipment specifications.

Process Hazard Analysis.

Based on the information gathered during the process safety information phase, the employer shall perform an initial process hazard analysis (hazard evaluation) on processes covered by this standard. The process hazard analysis shall be appropriate to the complexity of the process and shall identify, evaluate, and control the hazards involved in the process. Employers shall determine and document the priority order for conducting process hazard analyses based on a rationale which includes such considerations as extent of the process hazards, number of potentially affected employees, age of the process, and operating history of the process. The process hazard analysis shall be conducted as soon as possible, but not later than the following schedule:

Process hazards analyses completed after May 26, 1987 which meet the requirements of this paragraph are acceptable as initial process hazards analyses. These process hazard analyses shall be updated and revalidated, based on their completion date, in accordance with paragraph (e)(6) of this standard.

The employer shall use one or more of the following methodologies that are appropriate to determine and evaluate the hazards of the process being analyzed.

What-If;

Checklist;

What-If/Checklist;

Hazard and Operability Study (HAZOP);

Failure Mode and Effects Analysis (FMEA);

Fault Tree Analysis; or

An appropriate equivalent methodology.

The process hazard analysis shall address:

The hazards of the process;

The identification of any previous incident which had a likely potential for catastrophic consequences in the workplace;

Engineering and administrative controls applicable to the hazards and their interrelationships such as appropriate application of detection methodologies to provide early warning of releases. (Acceptable detection methods might include process monitoring and control instrumentation with alarms, and detection hardware such as hydrocarbon sensors.);

Consequences of failure of engineering and administrative controls; Facility siting;

Human factors; and

A qualitative evaluation of a range of the possible safety and health effects of failure of controls on employees in the workplace.

The process hazard analysis shall be performed by a team with expertise in engineering and process operations, and the team shall include at least one employee who has experience and knowledge specific to the process being evaluated. Also, one member of the team must be knowledgeable in the specific process hazard analysis methodology being used.

The employer shall establish a system to promptly address the team's findings and recommendations; assure that the recommendations are resolved in a timely manner and that the resolution is documented; document what actions are to be taken; complete actions as soon as possible; develop a written schedule of when these actions are to be completed; communicate the actions to operating, maintenance and other employees whose work assignments are in the process and who may be affected by the recommendations or actions.

At least every five (5) years after the completion of the initial process hazard analysis, the process hazard analysis shall be updated and revalidated by a team meeting the requirements, to assure that the process hazard analysis is consistent with the current process.

Employers shall retain process hazards analyses and updates or revalidations for each process covered by this section, as well as the documented resolution of recommendations, for the life of the process.

Operating Procedures.

The employer shall develop and implement written operating procedures that provide clear instructions for safely conducting activities involved in each covered process consistent with the process safety information and shall address at least the following elements.

Steps for each operating phase:

Initial startup;

Normal operations;

Temporary operations;

Emergency shutdown including the conditions under which emergency shutdown is required, and the assignment of shutdown responsibility to qualified operators to ensure that emergency shutdown is executed in a safe and timely manner.

Emergency Operations;

Normal shutdown; and,

Startup following a turnaround, or after an emergency shutdown.

Operating limits:

Consequences of deviation; and

Steps required to correct or avoid deviation.

Safety and health considerations:

Properties of, and hazards presented by, the chemicals used in the process;

Precautions necessary to prevent exposure, including engineering controls, administrative controls, and personal protective equipment;

Control measures to be taken if physical contact or airborne exposure occurs;

Quality control for raw materials and control of hazardous chemical inventory levels; and,

Any special or unique hazards.

Safety systems and their functions.

Operating procedures shall be readily accessible to employees who work in or maintain a process.

The operating procedures shall be reviewed as often as necessary to assure that they reflect current operating practice, including changes that result from changes in process chemicals, technology, and equipment, and changes to facilities. The employer shall certify annually that these operating procedures are current and accurate.

The employer shall develop and implement safe work practices to provide for the control of hazards during operations such as lockout/tagout; confined space entry; opening process equipment or piping; and control over entrance into a facility by maintenance, contractor, laboratory, or other support personnel. These safe work practices shall apply to employees and contractor employees.

Training

Each process must be operated consistently to insure that it operates safely. In order to insure consistent operation of the process, all personnel associated with the process must be trained. The training is mandatory and regulated. The following are some of the regulations impacting the training.

Initial training.

Each employee presently involved in operating a process, and each employee before being involved in operating a newly assigned process, shall be trained in an overview of the process and in the operating procedures as specified in paragraph (f) of this section. The training shall include emphasis on the specific safety and health hazards, emergency operations including shutdown, and safe work practices applicable to the employee's job tasks.

In lieu of initial training for those employees already involved in operating a process on May 26, 1992, an employer may certify in writing that the employee has the required knowledge, skills, and abilities to safely carry out the duties and responsibilities as specified in the operating procedures.

Refresher training.

Refresher training shall be provided at least every three years, and more often if necessary, to each employee involved in operating a process to assure that the employee understands and adheres to the current operating procedures of the process. The employer, in consultation with the employees involved in operating the process, shall determine the appropriate frequency of refresher training.

Training documentation.

The employer shall ascertain that each employee involved in operating a process has received and understood the training required by this paragraph. The employer shall prepare a record which contains the identity of the employee, the date of training, and the means used to verify that the employee understood the training.

Contractors

In many plants today, contract maintenance is becoming commonplace. However, just because the individuals working on the equipment are not employees of the company, the regulations do not allow for neglect of the documentation and training required for compliance. The following material highlights the responsibilities for the company and the contractors.

This section applies to contractors performing maintenance or repair, turnaround, major renovation, or specialty work on or adjacent to a covered process. It does not apply to contractors providing incidental services which do not influence process safety, such as janitorial work, food and drink services, laundry, delivery or other supply services.

Employer responsibilities.

The employer, when selecting a contractor, shall obtain and evaluate information regarding the contract employer's safety performance and programs.

The employer shall inform contract employers of the known potential fire, explosion, or toxic release hazards related to the contractor's work and the process.

The employer shall explain to contract employers the applicable provisions of the emergency action plan required by paragraph (n) of this section.

The employer shall develop and implement safe work practices consistent with paragraph (f)(4) of this section, to control the entrance, presence and exit of contract employers and contract employees in covered process areas.

The employer shall periodically evaluate the performance of contract employers in fulfilling their obligations as specified in paragraph (h)(3) of this section.

The employer shall maintain a contract employee injury and illness log related to the contractor's work in process areas.

Contract employer responsibilities.

The contract employer shall assure that each contract employee is trained in the work practices necessary to safely perform his/her job.

The contract employer shall assure that each contract employee is instructed in the known potential fire, explosion, or toxic release hazards related to his/her job and the process, and the applicable provisions of the emergency action plan.

The contract employer shall document that each contract employee has received and understood the training required by this paragraph. The contract employer shall prepare a record which contains the identity of the contract employee, the date of training, and the means used to verify that the employee understood the training.

The contract employer shall assure that each contract employee follows the safety rules of the facility including the safe work practices required by paragraph of this section.

The contract employer shall advise the employer of any unique hazards presented by the contract employer's work, or of any hazards found by the contract employer's work.

Pre-startup Safety Review

The pre-startup review is conducted prior to the commissioning of the equipment. This review is typically conducted by the engineering department or in progressive companies a team composed of representatives from engineering, maintenance, and operations. The review is conducted to determine the impact that the new equipment has on the PSM requirements. The regulations are very specific as to the data that must be recorded.

> The employer shall perform a pre-startup safety review for new facilities and for modified facilities when the modification is significant enough to require a change in the process safety information.
>
> The pre-startup safety review shall confirm that prior to the introduction of highly hazardous chemicals to a process:
>
>> Construction and equipment is in accordance with design specifications;
>>
>> Safety, operating, maintenance, and emergency procedures are in place and are adequate;
>>
>> For new facilities, a process hazard analysis has been per formed and recommendations have been resolved or implemented before startup; and modified facilities meet the requirements contained in management of change, paragraph (l).
>
> Training of each employee involved in operating a process has been completed.

Mechanical Integrity

This section of the regulations deals with insuring the condition of the process equipment. The regulations include documenting that the equipment was inspected, who inspected it, and the condition of the equipment at the time of the inspection. This again will highlight the importance of good documentation for the work order and preventive maintenance function to insure compliance.

Application.

> This section applies to the following process equipment:
>
> Pressure vessels and storage tanks;

Piping systems (including piping components such as valves);

Relief and vent systems and devices;

Emergency shutdown systems;

Controls (including monitoring devices and sensors, alarms, and interlocks) and,

Pumps.

Written procedures.

The employer shall establish and implement written procedures to maintain the on-going integrity of process equipment.

Training for process maintenance activities.

The employer shall train each employee involved in maintaining the on-going integrity of process equipment in an overview of that process and its hazards and in the procedures applicable to the employee's job tasks to assure that the employee can perform the job tasks in a safe manner.

Inspection and testing.

Inspections and tests shall be performed on process equipment.

Inspection and testing procedures shall follow recognized and generally accepted good engineering practices.

The frequency of inspections and tests of process equipment shall be consistent with applicable manufacturers' recommendations and good engineering practices, and more frequently if determined to be necessary by prior operating experience.

The employer shall document each inspection and test that has been performed on process equipment. The documentation shall identify the date of the inspection or test, the name of the person who performed the inspection or test, the serial number or other identifier of the equipment on which the inspection or test was performed, a description of the inspection or test performed, and the results of the inspection or test.

Equipment deficiencies.

The employer shall correct deficiencies in equipment that are outside acceptable limits (defined by the process safety information in paragraph (d) of this section) before further use or in a safe and timely manner when necessary means are taken to assure safe operation.

Quality assurance.

In the construction of new plants and equipment, the employer shall assure that equipment as it is fabricated is suitable for the process application for which they will be used.

Appropriate checks and inspections shall be performed to assure that equipment is installed properly and consistent with design specifications and the manufacturer's instructions.

The employer shall assure that maintenance materials, spare parts and equipment are suitable for the process application for which they will be used.

Hot Work Permit

Hot work permits are issued for repairs made on equipment and complete lockout/ tagout procedure can not be implemented. All safety precautions must be taken and proper notifications given to the employees performing the work and maintained during the work.

The employer shall issue a hot work permit for hot work operations conducted on or near a covered process.

The permit shall document that the fire prevention and protection requirements in 29 CFR 1910.252(a) have been implemented prior to beginning the hot work operations; it shall indicate the date(s) authorized for hot work; and identify the object on which hot work is to be performed. The permit shall be kept on file until completion of the hot work operations.

Management of Change

Since many of the processes covered in the process safety management regulations are hazardous, it is necessary to insure their continued safe operation. However, as equipment ages, vendors go out of business, and spare parts are discontinued, it may be necessary to find alternative spares. This section of the regulations addresses the procedures necessary to manage the change out of spare parts.

> The employer shall establish and implement written procedures to manage changes (except for "replacements in kind") to process chemicals, technology, equipment, and procedures; and, changes to facilities that affect a covered process.
>
> The procedures shall assure that the following considerations are addressed prior to any change:
>
> > The technical basis for the proposed change;
> >
> > Impact of change on safety and health;
> >
> > Modifications to operating procedures;
> >
> > Necessary time period for the change; and,
> >
> > Authorization requirements for the proposed change.
>
> Employees involved in operating a process and maintenance and contract employees whose job tasks will be affected by a change in the process shall be informed of, and trained in, the change prior to start-up of the process or affected part of the process.
>
> If a change covered by this paragraph results in a change in the process safety information required by paragraph (d) of this section, such information shall be updated accordingly.
>
> If a change covered by this paragraph results in a change in the operating procedures or practices required by paragraph (f) of this section, such procedures or practices shall be updated accordingly.

Incident Investigation.

Again, due to the hazardous nature of many of the processes, there is always a chance that something will cause a malfunction of the process equipment. If any hazardous chemical, or the potential release of a hazardous chemical, an incident report must be completed. This section details the documentation necessary to comply.

The employer shall investigate each incident which resulted in, or could reasonably have resulted in a catastrophic release of highly hazardous chemical in the workplace.

An incident investigation shall be initiated as promptly as possible, but not later than 48 hours following the incident.

An incident investigation team shall be established and consist of at least one person knowledgeable in the process involved, including a contract employee if the incident involved work of the contractor, and other persons with appropriate knowledge and experience to thoroughly investigate and analyze the incident.

A report shall be prepared at the conclusion of the investigation which includes at a minimum:

Date of incident;

Date investigation began;

A description of the incident;

The factors that contributed to the incident; and,

Any recommendations resulting from the investigation.

The employer shall establish a system to promptly address and resolve the incident report findings and recommendations. Resolutions and corrective actions shall be documented.

The report shall be reviewed with all affected personnel whose job tasks are relevant to the incident findings including contract employees where applicable.

Incident investigation reports shall be retained for five years.

Emergency Planning and Response.

Since the potential for a hazardous release always exists with any process, emergency planning to cope with the potential release is necessary. This section details the documentation necessary to insure the plan meets the requirements of the standards.

> The employer shall establish and implement an emergency action plan for the entire plant in accordance with the provisions of 29 CFR 1910.38(a). In addition, the emergency action plan shall include procedures for handling small releases. Employers covered under this standard may also be subject to the hazardous waste and emergency response provisions contained in 29 CFR 1910.120(a), (p) and (q).

Compliance Audits.

The regulations require periodic auditing of the practices and procedures developed by the company. This is to insure that the procedures are current and they are being followed to the correct level of detail. This section addresses the requirements for the compliance audit.

> Employers shall certify that they have evaluated compliance with the provisions of this section at least every three years to verify that the procedures and practices developed under the standard are adequate and are being followed.
>
> The compliance audit shall be conducted by at least one person knowledgeable in the process.
>
> A report of the findings of the audit shall be developed.
>
> The employer shall promptly determine and document an appropriate response to each of the findings of the compliance audit, and document that deficiencies have been corrected.
>
> Employers shall retain the two (2) most recent compliance audit reports.

Installation and Equipment Requirements

There are many equipment items, that while not a part of the process, provides utilizes to the process. The following section examines the regulations related to air systems. Note the requirements that are specified. While many of these items are considered good maintenance practices, they are rarely followed in many plants.

Installation.

Air receivers shall be so installed that all drains, hand holes, and manholes therein are easily accessible. Under no circumstances shall an air receiver be buried underground or located in an inaccessible place.

Drains and traps.

A drain pipe and valve shall be installed at the lowest point of every air receiver to provide for the removal of accumulated oil and water. Adequate automatic traps may be installed in addition to drain valves. The drain valve on the air receiver shall be opened and the receiver completely drained frequently and at such intervals as to prevent the accumulation of excessive amounts of liquid in the receiver.

Gages and valves.

Every air receiver shall be equipped with an indicating pressure gage (so located as to be readily visible) and with one or more spring-loaded safety valves. The total relieving capacity of such safety valves shall be such as to prevent pressure in the receiver from exceeding the maximum allowable working pressure of the receiver by more than 10 %.

No valve of any type shall be placed between the air receiver and its safety valve or valves.

Safety appliances, such as safety valves, indicating devices and controlling devices, shall be constructed, located, and installed so that they cannot be readily rendered inoperative by any means, including the elements.

All safety valves shall be tested frequently and at regular intervals to determine whether they are in good operating condition.

Use of mechanical equipment.

The same level of detail is also described for mechanical systems that support the process. The following regulations highlight this.

Where mechanical handling equipment is used, sufficient safe clearances shall be allowed for aisles, at loading docks, through doorways and wherever turns or passage must be made. Aisles and passageways shall be kept clear and in good repair, with no obstruction across or in aisles that could create a hazard. Permanent aisles and passageways shall be appropriately marked.

Secure storage.

Storage of material shall not createa hazard. Bags, containers, bundles, etc., stored in tiers shall be stacked, blocked, interlocked and limited in height so that they are stable and secure against sliding or collapse.

Housekeeping.

Storage areas shall be kept free from accumulation of materials that constitute hazards from tripping, fire, explosion, or pest harborage. Vegetation control will be exercised when necessary.

The user shall see that all nameplates and markings are in place and are maintained in a legible condition

Power Presses

One major item that cause many injuries each year are presses. There are many regulations that deal with presses. These include a full range of activities, from installation, operation and maintenance. The regulations are very specific, as can be seen in the following material.

Reconstruction and modification. It shall be the responsibility of any person reconstructing, or modifying a mechanical power press to do so in accordance with paragraph (b) of this section.

Excluded machines.

Press brakes, hydraulic and pneumatic power presses, bulldozers, hot bending and hot metal presses, forging presses and hammers, riveting machines and similar types of fastener applicators are excluded from the requirements of this section.

Mechanical Power Press Guarding and Construction, General

Hazards to personnel associated with broken or falling machine components. Machine components shall be designed, secured, or covered to minimize hazards caused by breakage, or loosening and falling or release of mechanical energy (i.e. broken springs).

Brakes.

Friction brakes provided for stopping or holding a slid movement shall be inherently self-engaging by requiring power or force from an external source to cause disengagement. Brake capacity shall be sufficient to stop the motion of the slide quickly and capable of holding the slide and its attachments at any point in its travel.

Machines using full revolution positive clutches.

Machines using full revolution clutches shall incorporate a single-stroke mechanism.

If the single-stroke mechanism is dependent upon spring action, the spring(s) shall be of the compression type, operating on a rod or guided within a hole or tube, and designed to prevent interleaving of the spring coils in event of breakage.

Foot pedals (treadle).

The pedal mechanism shall be protected to prevent unintended operation from falling or moving objects or by accidental stepping onto the pedal.

A pad with a nonslip contact area shall be firmly attached to the pedal.

The pedal return spring(s) shall be of the compression type, operating on a rod or guided within a hole or tube, or designed to prevent interleaving of spring coils in event of breakage.

If pedal counterweights are provided, the path of the travel of the weight shall be enclosed.

Hand operated levers.

Hand-lever-operated power presses shall be equipped with a spring latch on the operating lever to prevent premature or accidental tripping.

The operating levers on hand-tripped presses having more than one operating station shall be interlocked to prevent the tripping of the press except by the "concurrent" use of all levers.

Two-hand trip.

A two-hand trip shall have the individual operator's hand controls protected against unintentional operation and have the individual operator's hand controls arranged by design and construction and/or separation to require the use of both hands to trip the press and use a control arrangement requiring concurrent operation of the individual operator's hand controls.

Two-hand trip systems on full revolution clutch machines shall incorporate an antirepeat feature.

If two-hand trip systems are used on multiple operator presses, each operator shall have a separate set of controls.

Machines using part revolution clutches.

The clutch shall release and the brake shall be applied when the external clutch engaging means is removed, deactivated, or deenergized.

A red color stop control shall be provided with the clutch/brake control system. Momentary operation of the stop control shall immediately deactivate the clutch and apply the brake. The stop control shall override any other control, and reactuation of the clutch shall require use of the operating (tripping) means which has been selected.

A means of selecting Off, "Inch," Single Stroke, and Continuous (when the continuous function is furnished) shall be supplied with the clutch/brake control to select type of operation of the press. Fixing of selection shall be by means capable of supervision by the employer.

The "Inch" operating means shall be designed to prevent exposure of the workers hands within the point of operation by:

Requiring the concurrent use of both hands to actuate the clutch, or

Being a single control protected against accidental actuation and so located that the worker cannot reach into the point of operation while operating the single control.

Two-hand controls for single stroke shall conform to the following requirements:

Each hand control shall be protected against unintended operation and arranged by design, construction, and/or separation so that the concurrent use of both hands is required to trip the press.

The control system shall be designed to permit an adjustment which will require concurrent pressure from both hands during the die closing portion of the stroke.

The control system shall incorporate an antirepeat feature.

The control systems shall be designed to require release of all operators' hand controls before an interrupted stroke can be resumed. This requirement pertains only to those single-stroke, two-hand controls manufactured and installed on or after August 31, 1971.

Controls for more than one operating station shall be designed to be activated and deactivated in complete sets of two operator's hand controls per operating station by means capable of being supervised by the employer. The clutch/brake control system shall be designed and constructed to prevent actuation of the clutch if all operating stations are bypassed.

Those clutch/brake control systems which contain both single and continuous functions shall be designed so that completion of continuous circuits may be supervised by the employer.

The initiation of continuous run shall require a prior action or decision by the operator in addition to the selection of Continuous on the stroking selector, before actuation of the operating means will result in continuous stroking.

If foot control is provided, the selection method between hand and foot control shall be separate from the stroking selector and shall be designed so that the selection may be supervised by the employer.

Foot operated tripping controls, if used, shall be protected so as to prevent operation from falling or moving objects, or from unintended operation by accidental stepping onto the foot control.

The control of air-clutch machines shall be designed to prevent a significant increase in the normal stopping time due to a failure within the operating value mechanism, and to inhibit further operation if such failure does occur. This requirement shall apply only to those clutch/brake air-valve controls manufactured and installed on or after August 31, 1971, but shall not apply to machines intended only for continuous, automatic feeding applications.

The clutch/brake control shall incorporate an automatic means to prevent initiation or continued activation of the Single Stroke or Continuous functions unless the press drive motor is energized and in the forward direction

The clutch/brake control shall automatically deactivate in event of failure of the power or pressure supply for the clutch engaging means. Reactivation of the clutch shall require restoration of normal supply and the use of the tripping mechanism(s).

The clutch/brake control shall automatically deactivate in event of failure of the counterbalance(s) air supply. Reactivation of the clutch shall require restoration of normal air supply and use of the tripping mechanism(s).

Selection of bar operation shall be by means capable of being supervised by the employer. A separate pushbutton shall be employed to activate the clutch, and the clutch shall be activated only if the driver motor is deenergized.

Chapter 7

Electrical

A main power disconnect switch capable of being locked only in the OFF position shall be provided with every power press control system.

The motor start button shall be protected against accidental operation.

All mechanical power press controls shall incorporate a type of drive motor starter that will disconnect the drive motor from the power source in event of control voltage or power source failure, and require operation of the motor start button to restart the motor when voltage conditions are restored to normal.

All a.c. control circuits and solenoid value coils shall be powered by not more than a nominal 120-volt a.c. supply obtained from a transformer with an isolated secondary. Higher voltages that may be necessary for operation of machine or control mechanisms shall be isolated from any control mechanism handled by the operator, but motor starters with integral Start-Stop buttons may utilize line voltage control. All d.c. control circuits shall be powered by not more than a nominal 240-volt d.c. supply isolated from any higher voltages.

All clutch/brake control electrical circuits shall be protected against the possibility of an accidental ground in the control circuit causing false operation of the press.

Point of Operation Guards

Every point of operation guard shall meet the following design, construction, application, and adjustment requirements:

It shall prevent entry of hands or fingers into the point of operation by reaching through, over, under or around the guard;

It shall conform to the maximum permissible openings of Table O-10;

It shall, in itself, create no pinch point between the guard and moving machine parts;

It shall utilize fasteners not readily removable by operator, so as to minimize the possibility of misuse or removal of essential parts;

It shall facilitate its inspection, and

It shall offer maximum visibility of the point of operation consistent with the other requirements.

The employer shall furnish and enforce the use of hand tools for freeing and removing stuck work or scrap pieces from the die, so that no employee need reach into the point of operation for such purposes.

Inspection, Maintenance, and Modification of Presses

Inspection and maintenance records.

It shall be the responsibility of the employer to establish and follow a program of periodic and regular inspections of his power presses to ensure that all their parts, auxiliary equipment, and safeguards are in a safe operating condition and adjustment. The employer shall maintain a certification record of inspections which includes the date of inspection, the signature of the person who performed the inspection and the serial number, or other identifier, of the power press that was inspected.

Each press shall be inspected and tested no less than weekly to determine the condition of the clutch/brake mechan ism, antirepeat feature and single stroke mechanism. Necessary maintenance or repair or both shall be performed and completed before the press is operated. These requirements do not apply to those presses which comply with paragraphs (b)(13) and (14) of this section. The employer shall maintain a certification record of inspections, tests and maintenance work which includes the date of the inspection, test or maintenance; the signature of the person who performed the inspection, test, or maintenance; and the serial number or other identifier of the press that was inspected, tested or maintained.

Modification.

It shall be the responsibility of any person modifying a power press to furnish instructions with the modification to estab lish new or changed guidelines for use and care of the power press so modified.

Training of maintenance personnel.

It shall be the responsibility of the employer to insure the original and continuing competence of personnel caring for, inspecting, and maintaining power presses.

Operation of Power Presses

Instruction to operators.

The employer shall train and instruct the operator in the safe method of work before starting work on any operation covered by this section. The employer shall insure by adequate supervision that correct operating procedures are being followed.

Work area.

The employer shall provide clearance between machines so that movement of one operator will not interfere with the work of another. Ample room for cleaning machines, handling material, work pieces, and scrap shall also be provided. All surrounding floors shall be kept in good condition and free from obstructions, grease, oil, and water.

Overloading.

The employer shall operate his presses within the tonnage and attachment weight ratings specified by the manufacturer.

Flywheels and bearings.

Presses whose designs incorporate flywheels running on journals on the crankshaft or back shaft, or bull gears running on journals mounted on the crankshaft, shall be inspected, lubricated, and maintained as provided in paragraph (h)(10) of this section to reduce the possibility of unintended and uncontrolled press strokes caused by bearing seizure.

Inspection and Maintenance

Any press equipped with presence sensing devices for use in PSDI, or for supplemental safeguarding on presses used in the PSDI mode, shall be equipped with a test rod of diameter specified by the presence sensing device manufacturer to represent the minimum object sensitivity of the sensing field. Instructions for use of the test rod shall be noted on a label affixed to the presence sensing device.

The following checks shall be made at the beginning of each shift and whenever a die change is made.

A check shall be performed using the test rod according to the presence sensing device manufacturer's instructions to determine that the presence sensing device used for PSDI is operational.

The safety distance shall be checked for compliance with (h)(9)(v) of this section.

A check shall be made to determine that all supplemental safeguarding is in place. Where presence sensing devices are used for supplemental safeguarding, a check for proper operation shall be performed using the test rod according to the presence sensing device manufacturer's instructions.

A check shall be made to assure that the barriers and/or supplemental presence sensing devices required by paragraph (h)(9)(ix) of this section are operating properly.

A system or visual check shall be made to verify correct counterbalance adjustment for die weight according to the press manufacturer's instructions, when a press is equipped with a slide counterbalance system.

When presses used in the PSDI mode have flywheel or bullgear running on crankshaft mounted journals and bearings, or a flywheel mounted on back shaft journals and bearings, periodic inspections following the press manufacturer's recommendations shall be made to ascertain that bearings are in good working order, and that automatic lubrication systems for these bearings (if automatic lubrication is provided) are supplying proper lubrication. On presses with provision for manual lubrication of flywheel or bullgear bearings, lubrication shall be provided according to the press manufacturer's recommendations.

Periodic inspections of clutch and brake mechanisms shall be performed to assure they are in proper operating condition. The press manufacturer's recommendations shall be followed.

When any check of the press, including those performed in accordance with the requirements of paragraphs (h)(10)(ii), (iii) or (iv) of this section, reveals a condition of noncompliance, improper adjustment, or failure, the press shall not be operated until the condition has been corrected by adjustment, replacement, or repair.

It shall be the responsibility of the employer to ensure the competence of personnel caring for, inspecting, and maintaining power presses equipped for PSDI operation, through initial and periodic training.

After a press has been equipped with a safety system whose design has been certified and validated in accordance with paragraph (h)(11)(i) of this section, the safety system installation shall be certified by the employer, and then shall be validated by an OSHA-recognized third-party validation organization to meet all applicable requirements of paragraphs (a) through (h) and Appendix A of this section.

At least annually thereafter, the safety system on a mechanical power press used in the PSDI mode shall be recertified by the employer and revalidated by an OSHA-recognized third-party validation organization to meet all applicable requirements of paragraphs (a) through (h) and Appendix A of this section. Any press whose safety system has not been recertified and revalidated within the preceding 12 months shall be removed from service in the PSDI mode until the safety system is recertified and revalidated.

A label shall be affixed to the press as part of each installation certification/validation and the most recent recertification/revalidation. The label shall indicate the press serial number, the minimum safety distance (Ds) required by paragraph (h)(9)(v) of this section, the fulfillment of design certification/validation, the employer's signed certification, the identification of the OSHA-recognized third-party validation organization, its signed validation, and the date the certification/validation and recertification/revalidation are issued.

Records of the installation certification and validation and the most recent recertification and revalidation shall be maintained for each safety system equipped press by the employer as long as the press is in use. The records shall include the manufacture and model number of each component and subsystem, the calculations of the safety distance as required by paragraph (h)(9)(v) of this section, and the stopping time measurements required by paragraph (h)(2)(ii) of this section. The most recent records shall be made available to OSHA upon request.

The employer shall notify the OSHA-recognized third-party validation organization within five days whenever a component or a subsystem of the safety system fails or modifications are made which may affect the safety of the system. The failure of a critical component shall necessitate the removal of the safety system from service until it is recertified and revalidated, except recertification by the employer without revalidation is permitted when a non-critical component or subsystem is replaced by one of the same manufacture and design as the original, or determined by the third-party validation organization to be equivalent by similarity analysis, as set forth in Appendix A.

The employer shall notify the OSHA-recognized third-party validation organization within five days of the occurrence of any point of operation injury while a press is used in the PSDI mode. This is in addition to the report of injury required by paragraph (g) of this section; however, a copy of that report may be used for this purpose.

Operator Training

The operator training required by paragraph (f)(2) of this section shall be provided to the employee before the employee initially operates the press and as needed to maintain competence, but not less than annually thereafter. It shall include instruction relative to the following items for presses used in the PSDI mode.

The manufacturer's recommended test procedures for checking operationofthepresencesensingdevice.Thisshallincludetheuseof the test rod required by paragraph (h)(10)(i) of this section.

The safety distance required.

The operation, function and performance of the PSDmode.

The requirements for hand tools that may be used in the PSDI mode.

The severe consequences that can result if he or she at-tempts to circumvent or by-pass any of the safe-guard or operating functions of the PSDI system.

The employer shall certify that employees have been trained by preparing a certification record which includes the identity of the person trained, the signature of the employer or the person who conducted the training, and the date the training was completed. The certification record shall be prepared at the completion of training and shall be maintained on file for the duration of the employee's employment. The certification record shall be made available upon request to the Assistant Secretary for Occupational Safety and Health.

Presence sensing device initiation (PSDI).

Inspection and Maintenance

It shall be the responsibility of the employer to maintain all forge shop equipment in a condition which will insure continued safe operation. This responsibility includes:

Establishing periodic and regular maintenance safety checks and keeping certification records of these inspections which include the date of inspection, the signature of the person who performed the inspection and the serial number, or other identifier, for the forging machine which was inspected.

Scheduling and recording the inspection of guards and point of operation protection devices at frequent and reg ular intervals. Recording of inspections shall be in the form of a certification record which includes the date the inspection was performed, the signature of the person who performed the inspection and the serial number, or other identifier, of the equipment inspected.

Training personnel for the proper inspection and maintenance of forging machinery and equipment.

All overhead parts shall be fastened or protected in such a manner that they will not fly off or fall in event of failure.

Belt Fasteners.

Belts which of necessity must be shifted by hand and belts within seven (7) feet of the floor or working platform which are not guarded in accordance with this section shall not be fastened with metal in any case, nor with any other fastening which by construction or wear will constitute an accident hazard.

Standard Guard-General Requirements

Materials.

Standard conditions shall be secured by the use of the following materials. Expanded metal, perforated or solid sheet metal, wire mesh on a frame of angle iron, or iron pipe securely fastened to floor or to frame of machine.

All metal should be free from burrs and sharp edges.

Methods of manufacture.

Expanded metal, sheet or perforated metal, and wire mesh shall be securely fastened to frame.

Approved materials.

Minimum requirements. The materials and dimensions specified in this paragraph shall apply to all guards, except horizontal overhead belts, rope, cable, or chain guards more than seven (7) feet above floor, or platform.

All guards shall be rigidly braced every three (3) feet or fractional part of their height to some fixed part of machinery or building structure. Where guard is exposed to contact with moving equipment additional strength may be necessary.

CLEAN AIR ACT

While the Process Safety Management (PSM) regulations are not duplicated in many of the other regulatory agencies, the requirements for studying the processes and documented the expected results from the process are found. The following sections, starting with the Clean Air Act illustrates this point.

Air Pollution Documentation

Tthe owner or operator of such facility demonstrates, that emissions from construction or operation of such facility will not cause, or contribute to, air pollution in excess of any:

(A) maximum allowable increase or maximum allowable concentration for any pollutant in any area to which this part applies more than one time per year,

(B) national ambient air quality standard in any air quality control region, or

(C) any other applicable emission standard or standard of performance under this chapter;

The proposed facility is subject to the best available control technology for each pollutant subject to regulation under this chapter emitted from, or which results from, such facility;

There has been an analysis of any air quality impacts projected for the area as a result of growth associated with such facility;

> The person who owns or operates, or proposes to own or operate, a major emitting facility for which a permit is required under this part agrees to conduct such monitoring as may be necessary to determine the effect which emissions from any such facility may have, or is having, on air quality in any area which may be affected by emissions from such source;

FDA

As with the OSHA and the EPA, the FDA also regulates the documentation that must be kept during the construction, operation, and maintenance phases of an equipment's life cycle. The following material highlights the various aspects of the required documentation.

Plant Construction and Design

This first section examines the regulations concerning the buildings at the facility. These regulations focus on insuring the building is adequate to produce a quality product, free of contamination.

> Plant buildings and structures shall be suitable in size, construction, and design to facilitate maintenance and sanitary operations for food-manufacturing purposes. The plant and facilities shall:
>
> (1) Provide sufficient space for such placement of equipment and storage of materials as is necessary for the maintenance of sanitary operations and the production of safe food.
>
> (2) Permit the taking of proper precautions to reduce the poten tial for contamination of food, food-contact surfaces, or food-packaging materials with microorganisms, chemicals, filth, or other extraneous material. The potential for contamination may be reduced by adequate food safety controls and operat ing practices or effective design, including the separation of operations in which contamination is likely to occur, by one or more of the following means: location, time, partition, air flow, enclosed systems, or other effective means.
>
> (3) Permit the taking of proper precautions to protect food in outdoor bulk fermentation vessels by any effective means, includ ing:

(i) Using protective coverings.

(ii) Controlling areas over and aroundthe vessels to eliminate harborages for pests.

(iii) Checking on a regular basis for pests and pest infestation.

(iv) Skimming the fermentation vessels, as necessary.

(4) Be constructed in such a manner that floors, walls, and ceilings may be adequately cleaned and kept clean and kept in good repair; insuring that drip or condensate from fixtures, ducts and pipes does not contaminate food, food-contact surfaces, or food-packaging materials; and that aisles or working spaces are provided between equipment and walls and are adequately unobstructed and of adequate width to permit employees to perform their duties and to protect against contaminating food or food-contact surfaces with clothing or personal contact.

(5) Provide adequate lighting in hand-washing areas, dressing and locker rooms, and toilet rooms and in all areas where food is examined, processed, or stored and where equipment or utensils are cleaned; and provide safety-type light bulbs, fixtures, skylights, or other glass suspended over exposed food in any step of preparation or otherwise protect against food contamination in case of glass breakage.

(6) Provide adequate ventilation or control equipment to minimize odors and vapors (including steam and noxious fumes) in areas where they may contaminate food; and locate and operate fans and other air-blowing equipment in a manner that minimizes the potential for contaminating food, food-packaging materials, and food-contact surfaces.

(7) Provide, where necessary, adequate screening or other protection against pests.

Sanitary facilities and controls

Each plant shall be equipped with adequate sanitary facilities and accommodations including, but not limited to:

(a) Water supply.

The water supply shall be sufficient for the operations intended and shall be derived from an adequate source. Any water that contacts food or food-contact surfaces shall be safe and of adequate sanitary quality. Running water at a suitable temperature, and under pressure as needed, shall be provided in all areas where required for the processing of food, for the cleaning of equipment, utensils, and food-packaging materials, or for employee sanitary facilities.

(b) Plumbing.

Plumbing shall be of adequate size and design and adequately installed and maintained to:

(1) Carry sufficient quantities of water to required loca tions throughout the plant.

(2) Properly convey sewage and liquid disposable waste from the plant.

(3) Avoid constituting a source of contamination to food, water supplies, equipment, or utensils or creating an unsanitary condition.

(4) Provide adequate floor drainage in all areas where floors are subject to flooding-type cleaning or where normal operations release or discharge water or other liuid waste on the floor.

(5) Provide that there is not backflow from, or cross-connec tion between, piping systems that discharge waste water or sewage and piping systems that carry water for food or food manufacturing.

(c) Sewage disposal.

Sewage disposal shall be made into an adequate sewerage system or disposed of through other adequate means.

(d) Toilet facilities.

Each plant shall provide its employees with adequate, readily accessible toilet facilities. Compliance with this requirement may be accomplished by:

(1) Maintaining the facilities in a sanitary condition.

(2) Keeping the facilities in good repair at all times.

(3) Providing self-closing doors.

(4) Providing doors that do not open into areas where food is exposed to airborne contamination, except where alternate means have been taken to protect against such contamination (such as double doors or positive air-flow systems).

(e) Hand-washing facilities.

Hand-washing facilities shall be adequate and convenient and be furnished with running water at a suitable temperature. Compliance with this requirement may be accomplished by providing:

(1) Hand-wasing and, where appropriate, hand-sanitizing facilities at each location in the plant where good sanitary practices require employees to wash and/or sanitize their hands.

(2) Effective hand-cleaning and sanitizing preparations.

(3) Sanitary towel service or suitable drying devices.

(4) Devices or fixtures, such as water control valves, so designed and constructed to protect against recontamination of clean, sanitized hands.

(5) Readily understandable signs directing employeeshandling unprotected food, unprotected food-packaging materials, of food-contact surfaces to wash and, where appropriate, sanitize their hands before they start work, after each absence from post of duty, and when their hands may have become soiled or con taminated. These signs may be posted in the processing room(s) and in all other areas where employees may handle such food, materials, or surfaces.

(6) Refuse receptacles that are constructed and maintained in a manner that protects against contamination of food.

(f) Rubbish and offal disposal.

Rubbish and any offal shall be so conveyed, stored, and disposed of as to minimize the development of odor, minimize the potential for the waste becoming an attractant and harborage or breeding place for pests, and protect against contamination of food, food-contact surfaces, water supplies, and ground surfaces.

Design and Construction Features

(a) Any building or buildings used in the manufacture, processing, packing, or holding of a drug product shall be of suitable size, construction and location to facilitate cleaning, maintenance, and proper operations.

(b) Any such building shall have adequate space for the orderly placement of equipment and materials to prevent mix-ups between different components, drug product containers, closures, labeling, in-process materials, or drug products, and to prevent contamination. The flow of components, drug product containers, closures, labeling, in-process materials, and drug products through the building or buildings shall be designed to prevent contamination.

(c) Operations shall be performed within specifically defined areas of adequate size. There shall be separate or defined areas for the firm's operations to prevent contamination or mixups as follows:

(1) Receipt, identification, storage, and withholding from use of components, drug product containers, closures, and labeling, pending the appropriate sampling, testing, or examination by the quality control unit before release for manufacturing or packaging;

(2) Holding rejected components, drug product containers, closures, and labeling before disposition;

(3) Storage of released components, drug product containers, closures, and labeling;

(4) Storage of in-process materials;

(5) Manufacturing and processing operations;

(6) Packaging and labeling operations;

(7) Quarantine storage before release of drug products;

(8) Storage of drug products after release;

(9) Control and laboratory operations;

(10) Aseptic processing, which includes as appropriate:

(i) Floors, walls, and ceilings of smooth, hard surfaces that are easily cleanable;

(ii) Temperature and humidity controls;

(iii) An air supply filtered through high-efficiency particulat air filters under positive pressure, regardless of whether flow is laminar or nonlaminar;

(iv) A system for monitoring environmental conditions;

(v) A system for cleaning and disinfecting the room and equipment to produce aseptic conditions;

(vi) A system for maintaining any equipment used to control the aseptic conditions.

(d) Operations relating to the manufacture, processing, and packing of penicillin shall be performed in facilities seprate from those use for other drug products for human use.

Lighting.

Adequate lighting shall be provided in all areas.

Ventilation, air filtration, air heating and cooling.

(a) Adequate ventilation shall be provided.

(b) Equipment for adequate control over air pressure, microorganisms, dust, humidity, and temperature shall be provided when appropriate for the manufacture, processing, packing, or holding of a drug product.

(c) Air filtration systems, including prefilters and particulate matter air filters, shall be used when appropriate on air supplies to production areas. If air is recirculated to production areas, measures shall be taken to control recirculation of dust from production. In areas where air contamination occurs during production, there shall be adequate exhaust systems or other systems adequate to control contaminants.

(d) Air-handling systems for the manufacture, processing, and packing of penicillin shall be completely separate from those for other drug products for human use.

Plumbing.

(a) Potable water shall be supplied under continuous positive pressure in a plumbing system free of defects that could contribute contamination to any drug product. Potable water shall meet the standards prescribed in the Environmental Protection Agency's Primary Drinking Water Regulations set forth in 40 CFR part 141. Water not meeting such standards shall not be permitted in the potable water system.

(b) Drains shall be of adequate size and, where connected directly to a sewer, shall be provided with an air break or other mechanical device to prevent back-siphonage.

Sewage and refuse.

Sewage, trash, and other refuse in and from the building and immediate premises shall be disposed of in a safe and sanitary manner.

Washing and toilet facilities.

Adequate washing facilities shall be provided, including hot and cold water, soap or detergent, air driers or single-service towels, and clean toilet facilities easily accessible to working areas.

Sanitation.

(a) Any building used in the manufacture, processing, packing, or holding of a drug product shall be maintained in a clean and sanitary condition, Any such building shall be free of infestation by rodents, birds, insects, and other vermin (other than laboratory animals). Trash and organic waste matter shall be held and disposed of in a timely and sanitary manner.

(b) There shall be written procedures assigning responsibility for sanitation and describing in sufficient detail the cleaning schedules, methods, equipment, and materials to be used in cleaning the buildings and facilities; such written procedures shall be followed.

(c) There shall be written procedures for use of suitable rodenticides, insecticides, fungicides, fumigating agents, and cleaning and sanitizing agents. Such written procedures shall be designed to prevent the contamination of equipment, components, drug product containers, closures, packaging, labeling materials, or drug products and shall be followed. Rodenticides, insecticides, and fungicides shall not be used unless registered and used in accordance with the Federal Insecticide, Fungicide, and Rodenticide Act (7 U.S.C. 135)

(d) Sanitation procedures shall apply to work performed by contractors or temporary employees as well as work performed by full-time employees during the ordinary course of operations.

Maintenance.

Any building used in the manufacture, processing, packing, or holding of a drug product shall be maintained in a good state of repair

Production Equipment

In addition to all of the facility regulations, there are specific regulations to consider during the design, installation, operation and maintenance phases of the equipment life cycle. These are considered in the following section.

Equipment design, size, and location.

Equipment used in the manufacture, processing, packing, or holding of a drug product shall be of appropriate design, adequate size, and suitably located to facilitate operations for its intended use and for its cleaning and maintenance.

Equipment construction.

(a) Equipment shall be constructed so that surfaces that contact components, in-process materials, or drug products shall not be reactive, additive, or absorptive so as to alter the safety, identity, strength, quality, or purity of the drug product beyond the official or other established requirements.

(b) Any substances required for operation, such as lubricants or coolants, shall not come into contact with components, drug product containers, closures, in-process materials,or drug products so as to alter the safety, identity, strength, quality, or purity of the drug product beyond the official or other established requirements.

Equipment cleaning and maintenance.

(a) Equipment and utensils shall be cleaned, maintained, and sanitized at appropriate intervals to prevent malfunctions or contamination that would alter the safety, identity, strength, quality, or purity of the drug product beyond the official or other established requirements.

(b) Written procedures shall be established and followed for cleaning and maintenance of equipment, including utensils, used in the manufacture, processing, packing, or holding of a drug product. These procedures shall include, but are not necessarily limited to, the following:

(1) Assignment of responsibility for cleaning and maintaining equipment;

(2) Maintenance and cleaning schedules, including, where appropriate, sanitizing schedules;

(3) A description in sufficient detail of the methods, equipment, and materials used in cleaning and maintenance operations, and the methods of disassembling and reassembling equipment as necessary to assure proper cleaning and maintenance;

(4) Removal or obliteration of previous batch identification;

(5) Protection of clean equipment from contamination prior to use;

(6) Inspection of equipment for cleanliness immediately before use.

Automatic, mechanical, and electronic equipment.

(a) Automatic, mechanical, or electronic equipment or other types of equipment, including computers, or related systems that will perform a function satisfactorily, may be used in the manufacture, processing, packing, and holding of a drug product. If such equipment is so used, it shall be routinely calibrated, inspected, or checked according to a written program designed to assure proper performance. Written records of those calibration checks and inspections shall be maintained.

(b) Appropriate controls shall be exercised over computer or related systems to assure that changes in master production and control records or other records are instituted only by authorized personnel. Input to and output from the computer or related system of formulas or other records or data shall be checked for accuracy. The degree and frequency of input/output verification shall be based on the complexity and reliability of the computer or related system. A backup file of data entered into the computer or related system shall be maintained except where certain data, such as calculations performed in connection with laboratory analysis, are eliminated by computerization or other automated processes. In such instances a written record of the program shall be maintained along with appropriate validation data. Hard copy or alternative systems, such as duplicates, tapes, or microfilm, designed to assure that backup data are exact and complete and that it is secure from alteration, inadvertent erasures, or loss shall be maintained.

Filters.

Filters for liquid filtration used in the manufacture, process ing, or packing of injectable drug products intended for human use shall not release fibers into such products. Fiber-releasing filters may not be used in the manufacture, process ing, or packing of these injectable drug products unless it is not possible to manufacture such drug products without the use of such filters. If use of a fiber-releasing filter is necessary, an additional non-fiber-releasing filter of 0.22 micron maximum mean porosity (0.45 micron if the manufacturing conditions so dictate) shall subsequently be used to reduce the content of particles in the injectable drug product. Use of an asbestos-containing filter, with or without subsequent use of a specific non-fiber-releasing filter, is permissible only upon submission of proof to the appropriate bureau of the Food and Drug Administration that use of a non-fiber-releasing filter will, or is likely to, compromise the safety or effectiveness of the injectable drug product.

Written procedures; deviations.

(a) There shall be written procedures for production and process control designed to assure that the drug products have the identity, strength, quality, and purity they purport or are represented to possess. Such procedures shall include all requirements in this subpart. These written procedures, including any changes, shall be drafted, reviewed, and approved by the appropriate organizational units and reviewed and approved by the quality control unit.

(b) Written production and process control procedures shall be followed in the execution of the various production and process control functions and shall be documented at the time of performance. Any deviation from the written procedures shall be recorded and justified.

Additional FDA Guidelines

In addition to the previous regulations, the following sections contain some guidelines specific to certain operations. These also have an impact on the design, operation, and maintenance of FDA regulated facilities.

Aseptic Facilities

Facility design for the aseptic processing of sterile bulk drug substances should have the same design features as an SVP aseptic processing facility. These would include temperature, humidity and pressure control. Because sterile bulk aseptic facilities are usually larger, problems with pressure differentials and sanitization have been encountered. For example, a manufacturer was found to have the gowning area under greater pressure than the adjacent aseptic areas. The need to remove solvent vapors may also impact on area pressurization.

Unnecessary equipment and/or equipment that cannot be adequately sanitized, such as wooden skids and forklift trucks, should be identified. Inquire about the movement of large quantities of sterile drug substance and the location of pass-through areas between the sterile core and non-sterile areas. Observe these areas, review environmental monitoring results and sanitization procedures.

The CGMP Regulations prohibit the use of asbestos filters in the final filtration of solutions. At present, it would be difficult for a manufacturer to justify the use of asbestos filters for filtration of air or solutions. Inquire about the use of asbestos filters.

Facilities used for the charge or addition of non-sterile components, such as the non-sterile drug substance, should be similar to those used for the compounding of parenteral solutions prior to sterilization. The concern is soluble extraneous contaminants, including endotoxins, that may be carried through the process. Observe this area and review the environmental controls and specifications to determine the viable and non-viable particulate levels allowed in this area.

Equipment.

Equipment used in the processing of sterile bulk drug substances should be sterile and capable of being sterilized. This includes the crystallizer, centrifuge and dryer. The sanitization, rather than sterilization of this equipment, is unacceptable. Sterilization procedures and the validation of the sterilization of suspect pieces of equipment and transfer lines should be reviewed.

The method of choice for the sterilization of equipment and transfer lines is saturated clean steam under pressure. In the validation of the sterilization of equipment and of transfer systems, Biological Indicators (BIs), as well as temperature sensors (Thermocouple (TC) or Resistance Thermal Device (RTD)) should be strategically located in cold spots where condensate may accumulate. These include the point of steam injection and steam discharge, as well as cold spots, which are usually low spots. For example, in a recent inspection, a manufacturer utilized a Sterilize-In-Place (SIP) system and only monitored the temperature at the point of discharge and not in low spots in the system where condensate can accumulate.

EPA Building Design Background

In considering the design requirements for buildings, the EPA has some interesting commentary about building design and indoor air quality. The following EPA commentary shows the impact that poor indoor air quality has on the personnel working inside the facility buildings. One of the most common problems is Sick Building Syndrome (SBS), where various building problems combine to create a working environment that impacts employee health.

Indoor Air Quality

Indoor air quality problems can occur in all types and ages of buildings; in newly constructed buildings, in renovated or remodeled buildings, and in old buildings. Problems in new, clean buildings are rarely, if ever, related to microbial growth, since the physical structures are new.

Older buildings that have not been adequately maintained and operated may have problems with bioaerosols if parts of the building have been allowed to become reservoirs for microbial growth. Also, if inadequate outside air is provided, regardless of the age of the building, chemical and biological contaminants will build up to levels that can cause health effects in some workers. In addition, other physical factors such as lack of windows, noise, and inadequate lighting, and ergonomic factors involving uncomfortable furniture and intensive use of video display units, etc., will cause discomfort in occupants that may be inaccurately attributed to air quality.

Cardiovascular effects have also been associated with poor indoor air quality. These effects are presented as headache, fatigue, dizziness, aggravation of existing cardiovascular disease, and damage to the heart. These effects are associated with exposure to combustion gases such as carbon monoxide Volatile Organic Compounds (VOCs) and particulates.

Nervous system effects have also been produced due to exposure to poor indoor air quality. These effects include headache, blurred vision, fatigue, malaise with nausea, ringing in the ears, impaired judgment, and polyneuritis. These effects are associated with exposure to carbon dioxide, carbon monoxide, formaldehyde, and VOCs.

Approximately 15% (20,250) of 135,000 hospital admissions per year that last an average of more than eight days are due to allergic disease. One study estimates estimate that these hospitalizations cost five million work days per year.

In some instances, outbreaks of SBS are identified with specific pollutant exposures, but in general only general etiologic factors related to building design, operation and maintenance can be identified

Many individuals who believe the building they work in is implicated in SBS have described similar effects. Symptoms usually include one or more of the following: mucous membrane (eye, nose, or throat) irritation, dry skin, headache, nausea, fatigue, and lethargy. These symptoms are generally believed to result from indoor air pollution.

Indoor air pollution may be caused by physical, chemical, or microbiological agents, and is aggravated by poor ventilation.

Note: the following refers to an EPA study of chemicals commonly used inside a building.

> Another study showed a number of problems ascribed to indoor air pollution in the chemically sensitive patient. These problems include irritability from natural gas fumes, allergy to dust from forced air ventilation systems, intoxication and even hallucination from paint fumes. Randolph describes chemical sensitivity to dry cleaning chemicals and rug shampoo, and implicates moldy carpets in producing allergenic substances. He also describes joint pain, malaise, and fatigue due to pesticide exposure; and skin rashes from exposure to plasticizers. Randolph further describes intolerance to highly scented products such as deodorant soaps, toilet deodorants, and disinfectants, especially pine-scented ones. Other patients have reported reacting to strong perfumes and other cosmetics. So-called air fresheners often prove to be particularly troublesome. He also describes that some patients are sensitive to the odors from hot plastic-coated wires in electronic equipment.

> In summary, sick building syndrome is not a well-defined disease with well-defined causes. It appears to be a reaction, at least in part due to stimulation of the common chemical sense, to a variety of chemical, physical or biological stimuli. Its victims display all or some of a pattern of irritation of the mucous membranes, and the worst affected individuals have neurological symptoms as well.

> Individuals with underlying pulmonary disease, such as asthma, are more susceptible than others to acute exposure to these indoor air contaminants and experience coughing and wheezing at low levels of exposure. Synergism may occur between chemical contaminants, such as ozone and VOCs, in aggravating asthma. These affected individuals may also be at increased risk of pulmonary infections due to the synergistic effect between chemical and microbial contaminants.

> Microbial contamination of building structures, furnishings, and HVAC system components contribute to poor indoor air quality problems, especially those related to building-related illnesses. OSHA believes that consequent health effects con-

stitute material impairment of health. These can be categorized as irritation, pulmonary, cardiovascular, nervous system, reproductive, and cancer effects.

In addition, water leakage on furnishings or within building components can result in the proliferation of microorganisms that can release acutely irritating substances into the air. Typically, where microorganisms are allowed to grow, a moldy smell develops. This moldy smell is often associated with microbial contamination and is a result of VOCs released during microbial growth on environmental substrates

As can be seen from the above reference material, there are many factors that must be considered in the design and maintenance of building structures. The study clearly shows that if a comprehensive study, considering the factors pointed out previously, that occupying buildings can have a damaging impact on employee health. This, in turn, will have a dramatic impact on the company's profitability.

CHAPTER 8

Maintenance Management and Regulatory Compliance

In industry today, companies fall into one of three categories:

(a) Those that try to meet the regulatory requirements

(b) Those that fail to meet the requirements for the lack of understanding

(c) Those that deliberately ignore the requirements

Meeting the Requirements

Organizations that meet the regulatory requirements have strong maintenance and operational practices and procedures in place. They clearly understand the regulations and make every effort to comply. They are managed by responsible and knowledgeable executives. To them, compliance issues are viewed as a necessary price to do business in their respective industries.

Lack of Understanding

Some organizations lack the technical knowledge to properly manage their businesses. They fail to understand that it is a combination of company assets (e.g., equipment, raw materials, facilities) and employees that provide their product or service to their respective customers. Because they lack the proper technical skills, these organizations tend to make bad decisions about their equipment or technical processes. The skills may, in fact, even exist somewhere within their organizations, but important information and perspectives never reach the key decision makers. These organizations will have regulatory violations for the wrong reasons.

Deliberately Ignoring the Requirements

While relatively few in number, there are organization that deliberate ignore regulatory requirements. Appendix C identifies repetitive violators. By deliberately disregarding the regulations, these organizations put the lives of both their employees and their customers at risk. The irresponsible managers who lead these organizations deserved to be removed from their positions and replaced with more responsible executives. When they are caught – and eventually they always are – they incur a tremendous financial liability for their companies. These are the most reprehensible of all managers found in companies today.

When deciding whether or not to comply with the regulations, two factors are important. First and foremost, the regulations are not optional. But second, there are financial reasons why compliance makes sense.

The Cost of Non-Compliance

Non-compliance has its costs. A quick look at the following OSHA web site — http://www.osha.gov/oshstats/std2.html — can demonstrate this point. This site lists fines levied by OSHA inspectors for the previous year. For 2001–2002, the total is a staggering $77,686,249.71. Furthermore, this figure, large in itself, does not even include EPA and FDA fines.

Note the amounts mentioned in the following excerpt from the EPA:

> Most States collect fines and penalties, and dedicate them to environmental programs. Many have set up trust funds to receive the payments and then spend monies for environmental purposes (includes Florida, Indiana, Massachusetts, Missouri, Montana, Ohio, Pennsylvania, New Jersey, New York, Virginia, Washington and Wisconsin). Most of these States collect several millions in fines and penalties annually, with New Jersey heading the list with almost $500 million in one year. For example, New York collected a $3 million criminal penalty in 1994 from one company, which also had to build a $20 million industrial pretreatment facility.

The following excerpts of an FDA case point to substantial fines that may be levied for non-compliance with regulations.:

> One major criminal case involved ... FDA investigators found that the company had for years been substituting colored water for apple juice. A 470-count indictment of the company and some of its officers was filed charging conspiracy, mail fraud, and marketing artificially flavored sugar water as apple juice concentrate with the intention to defraud and mislead, a felony violation of the FD&C Act. The company pleaded guilty to 215 felony violations of the FD&C Act and was sentenced to pay fines and costs totaling almost $2.2 million. The vice president for operations was tried, convicted and sentenced to a year and a day in jail, and the president of the company to six months of community service.
>
> In addition, each had to pay a $100,000 fine. Five companies that supplied raw materials to...entered into plea-bargaining and received lesser sentences.

Any ISO violations or non-conformance typically results in the biggest penalty: lost customers.

Note that Appendix C lists only a sample of equipment-related OSHA violations. If all the OSHA, EPA and FDA listings were included, this book would be thousands of pages longer.

The Cost of Compliance

What is the cost of compliance? Much of it is the cost of a good maintenance management program, including:

- Preventive maintenance
- Inventory and purchasing
- Work order systems, including planning and scheduling
- Computerized Maintenance Management Systems
- Training and education
- Operations involvement
- Reliability Centered Maintenance?

Unfortunately, many companies today struggle with the basics. They fail to understand that good maintenance initiatives pay for themselves many times over. They start a good maintenance program, that is fully compliant with all of the regulatory agencies. However, in a business downturn, they downsize the maintenance organization, removing the resources necessary to maintain compliance. In reality, the costs to maintain compliance are not expensive. So the company is actually trying to meet some "headcount" parameter and not a true financial target.

Some of the specific costs of compliance include:

Labor

Materials

Recordkeeping systems

The relevant labor costs are simply those costs involved in carrying out the activities needed to meet regulatory requirements. This would include the labor to physically perform the work and the labor to record the necessary documentation. The materials cost would include the spare parts that are changed on specific predetermined frequencies, instead of the run to failure schedule that is used in most companies today. The recordkeeping system costs include the computers and software that would be required to comply with the regulations.

To estimate compliance costs, use the following steps: :

1. Specify the labor and materials requirements to perform all of the regulatory maintenance tasks (typically PMs)
2. Calculate the hours required for each PM
3. Calculate the material cost for each PM
4. Determine the frequency of each PM
   Number of times per year to be performed
5. Multiply costs X frequency
6. Total the results to find the cost of labor and material compliance
7. Evaluate CMMS packages to determine which ones meet the information needs required to comply

8. Price the system, hardware and training Also calculate the labor necessary to gather and maintain on a daily basis the data required for compliance
9. Total the results from steps 7 and 8 to find the cost of record keeping.
10. Total the costs from steps 6 and 9 to find the compliance cost.

From a maintenance perspective, similar steps can be used to determine manpower loading. Using the data from above, calculate the total hours for compliance from a maintenance department perspective. This figure represents the necessary labor total for the maintenance labor. It also becomes the baseline. Any other maintenance activities require resources beyond this baseline. These resources should not be compromised; otherwise, the regulatory compliance for the company will be compromised.

It is clear, that the cost of violating any of the regulatory requirements is beyond what the cost of comply with the same requirements would be.

THE FUTURE

In the future maintenance management will become an integral part of all regulatory compliance strategies. Without an effective maintenance organization, a company can not possibly meet its regulatory obligations. However, there is a major problem inside most corporations today, which is the lack of understand of maintenance/ asset management. Without a clear understand of all of the facets of a business that is impacted, managers make many incorrect decisions about this part of their business.

Only when maintenance/ asset management becomes an integral part of the curriculum at business schools will this problem be resolved. It is hoped that by highlighting the impact that maintenance/ asset management has on regulatory compliance, another step is taken in that direction.

This text should prove useful to organizations desiring to achieve regulatory compliance and ultimately succeed in their respective marketplaces.

Introduction to Appendices

The Appendices that follow are included to highlight the detail involved in maintaining compliance with regulatory agencies. The samples that are included highlight the need for adequate personnel that are properly trained and educated to help a comply maintain compliance with the standards.

Appendix A
OSHA Standards and Related Permit Forms

PART 1910 - OCCUPATIONAL SAFETY AND HEALTH STANDARDS

Subpart A - General
Sec.
1910.1 Purpose and scope.
1910.2 Definitions.
1910.3 Petitions for the issuance, amendment, or repeal of a standard.
1910.4 Amendments to this part.
1910.5 Applicability of standards.
1910.6 Incorporation by reference.
1910.7 Definition and requirements for a nationally recognized testing laboratory.
1910.8 OMB control numbers under the Paperwork Reduction Act.

Subpart B - Adoption and Extension of Established Federal Standards

1910.11 Scope and purpose.
1910.12 Construction work.
1910.15 Shipyard employment.
1910.16 Longshoring and marine terminals.
1910.17 Effective dates.
1910.18 Changes in established Federal standards.
1910.19 Special provisions for air contaminants.

Appendix A

Subpart C - [Removed and Reserved]

1910.20 [Redesignated as 1910.1020]

Subpart D - Walking - Working Surfaces

1910.21 Definitions.
1910.22 General requirements.
1910.23 Guarding floor and wall openings and holes.
1910.24 Fixed industrial stairs.
1910.25 Portable wood ladders.
1910.26 Portable metal ladders.
1910.27 Fixed ladders.
1910.28 Safety requirements for scaffolding.
1910.29 Manually propelled mobile ladder stands and scaffolds (towers).
1910.30 Other working surfaces.

Subpart E - Means of Egress

1910.35 Definitions.
1910.36 General requirements.
1910.37 Means of egress, general.
1910.38 Employee emergency plans and fire prevention plans.

APPENDIX TO SUBPART E - MEANS OF EGRESS

Subpart F - Powered Platforms, Manlifts, and Vehicle-Mounted Work Platforms.

1910.66 Powered platforms for building maintenance.
1910.67 Vehicle-mounted elevating and rotating work platforms.
1910.68 Manlifts.

Subpart G - Occupational Health and Environmental Control

1910.94 Ventilation.
1910.95 Occupational noise exposure.
1910.96 [Redesignated as 1910.1096]
1910.97 Nonionizing radiation.
1910.98 Effective dates.

Subpart H - Hazardous Materials

1910.101 Compressed gases (general requirements).
1910.102 Acetylene.
1910.103 Hydrogen.

1910.104 Oxygen.
1910.105 Nitrous oxide.
1910.106 Flammable and combustible liquids.
1910.107 Spray finishing using flammable and combustible materials.
1910.108 Dip tanks containing flammable or combustible liquids.
1910.109 Explosives and blasting agents.
1910.110 Storage and handling of liquified petroleum gases.
1910.111 Storage and handling of anhydrous ammonia.
1910.112 [Reserved]
1910.113 [Reserved]
1910.119 Process safety management of highly hazardous chemicals.
1910.120 Hazardous waste operations and emergency response.
1910.121 [Reserved]
1910.122 Table of contents.
1910.123 Dipping and coating operations: Coverage and Definitions.
1910.124 General requirements for dipping and coating operations.
1910.125 Additional requirements for dipping and coating operations that use flammable or combustible liquids.
1910.126 Additional requirements for special dipping and coating applications.

Subpart I - Personal Protective Equipment

1910.132 General requirements.
1910.133 Eye and face protection.
1910.134 Respiratory protection.
1910.135 Head protection.
1910.136 Foot protection.
1910.137 Electrical protective devices.
1910.138 Hand Protection.
1910.139 Respiratory protection for M. tuberculosis.

Subpart J - General Environmental Controls

1910.141 Sanitation.
1910.142 Temporary labor camps.
1910.143 Nonwater carriage disposal systems. [Reserved]
1910.144 Safety color code for marking physical hazards.
1910.145 Specifications for accident prevention signs and tags.
1910.146 Permit-required confined spaces.
1910.147 The control of hazardous energy (lockout/tagout).

Subpart K - Medical and First Aid

1910.151 Medical services and first aid.
1910.152 [Reserved]

Appendix A

Subpart L - Fire Protection

1910.155 Scope, application and definitions applicable to this subpart.
1910.156 Fire brigades.

PORTABLE FIRE SUPPRESSION EQUIPMENT

1910.157 Portable fire extinguishers.
1910.158 Standpipe and hose systems.

FIXED FIRE SUPPRESSION EQUIPMENT

1910.159 Automatic sprinkler systems.
1910.160 Fixed extinguishing systems, general.
1910.161 Fixed extinguishing systems, dry chemical.
1910.162 Fixed extinguishing systems, gaseous agent.
1910.163 Fixed extinguishing systems, water spray and foam.

OTHER FIRE PROTECTIVE SYSTEMS

1910.164 Fire detection systems.
1910.165 Employee alarm systems.

APPENDICES TO SUBPART L

APPENDIX A TO SUBPART L - FIRE PROTECTION
APPENDIX B TO SUBPART L - NATIONAL CONCENSUS STANDARDS
APPENDIX C TO SUBPART L - FIRE PROTECTION REFERENCES FOR FURTHER INFORMATION
APPENDIX D TO SUBPART L - AVAILABILITY OF PUBLICATIONS INCORPORATED BY REFERENCE IN SECTION 1910.156 FIRE BRIGADES
APPENDIX E TO SUBPART L - TEST METHODS FOR PROTECTIVE CLOTHING

Subpart M - Compressed Gas and Compressed Air Equipment

1910.166 [Reserved]
1910.167 [Reserved]
1910.168 [Reserved]
1910.169 Air receivers.

Subpart N - Materials Handling and Storage

1910.176 Handling material - general.

1910.177 Servicing multi-piece and single piece rim wheels.
1910.178 Powered industrial trucks.
1910.179 Overhead and gantry cranes.
1910.180 Crawler locomotive and truck cranes.
1910.181 Derricks.
1910.183 Helicopters.
1910.184 Slings.

APPENDIX A to 1910.178 - Stability of Powered Industrial Trucks (non-mandatory Appendix to Paragraph (l) of this section.

Subpart O - Machinery and Machine Guarding

1910.211 Definitions.
1910.212 General requirements for all machines.
1910.213 Woodworking machinery requirements.
1910.214 Cooperage machinery.
1910.215 Abrasive wheel machinery.
1910.216 Mills and calenders in the rubber and plastics industries.
1910.217 Mechanical power presses.
1910.218 Forging machines.
1910.219 Mechanical power-transmission apparatus.

Subpart P - Hand and Portable Powered Tools and Other Hand-Held Equipment.

1910.241 Definitions.
1910.242 Hand and portable powered tools and equipment, general.
1910.243 Guarding of portable powered tools.
1910.244 Other portable tools and equipment.

Subpart Q - Welding, Cutting, and Brazing.

1910.251 Definitions.
1910.252 General requirements.
1910.253 Oxygen-fuel gas welding and cutting.
1910.254 Arc welding and cutting.
1910.255 Resistance welding.

Subpart R - Special Industries

1910.261 Pulp, paper, and paperboard mills.
1910.262 Textiles.
1910.263 Bakery equipment.
1910.264 Laundry machinery and operations.
1910.265 Sawmills.

1910.266 Logging operations.
1910.267 [Reserved]
1910.268 Telecommunications.
1910.269 Electric power generation, transmission, and distribution.
1910.272 Grain handling facilities.

Subpart S - Electrical

GENERAL

1910.301 Introduction.

DESIGN SAFETY STANDARDS FOR ELECTRICAL SYSTEMS

1910.302 Electric utilization systems.
1910.303 General requirements.
1910.304 Wiring design and protection.
1910.305 Wiring methods, components, and equipment for general use.
1910.306 Specific purpose equipment and installations.
1910.307 Hazardous (classified) locations.
1910.308 Special systems.
1910.309 - 1910.330 [Reserved]

SAFETY-RELATED WORK PRACTICES

1910.331 Scope.
1910.332 Training.
1910.333 Selection and use of work practices.
1910.334 Use of equipment.
1910.335 Safeguards for personnel protection.
1910.336 - 1910.360 [Reserved]

SAFETY-RELATED MAINTENANCE REQUIREMENTS

1910.361 - 1910.380 [Reserved]

SAFETY REQUIREMENTS FOR SPECIAL EQUIPMENT

1910.381 - 1910.398 [Reserved]

DEFINITIONS

1910.399 Definitions applicable to this subpart.

APPENDIX A TO SUBPART S - REFERENCE DOCUMENTS
APPENDIX B TO SUBPART S - EXPLANATORY DATA [RESERVED]
APPENDIX C TO SUBPART S - TABLES, NOTES, AND CHARTS[Reserved

Appendix A

Subpart T - Commercial Diving Operations

GENERAL

1910.401 Scope and application.
1910.402 Definitions.

PERSONNEL REQUIREMENTS

1910.410 Qualifications of dive team.

GENERAL OPERATIONS PROCEDURES

1910.420 Safe practices manual.
1910.421 Pre-dive procedures.
1910.422 Procedures during dive.
1910.423 Post-dive procedures.

SPECIFIC OPERATIONS PROCEDURES

1910.424 SCUBA diving.
1910.425 Surface-supplied air diving.
1910.426 Mixed-gas diving.
1910.427 Liveboating.

EQUIPMENT PROCEDURES AND REQUIREMENTS

1910.430 Equipment.

RECORDKEEPING

1910.440 Recordkeeping requirements.
1910.441 Effective date.

APPENDIX A TO SUBPART T - EXAMPLES OF CONDITIONS WHICH MAY RESTRICT OR LIMIT EXPOSURE TO HYPERBARIC CONDITIONS
APPENDIX B TO SUBPART T - GUIDELINES FOR SCIENTIFIC DIVING

Subparts U - Y [Reserved]

1910.442 - 1910.999 [Reserved]

Subpart Z - Toxic and Hazardous Substances

1910.1000 Air contaminants.

Appendix A

1910.1001 Asbestos.
1910.1002 Coal tar pitch volatiles; interpretation of term.
1910.1003 13 Carcinogens (4-Nitrobiphenyl, etc.).
1910.1004 alpha-Naphthylamine.
1910.1005 [Reserved]
1910.1006 Methyl chloromethyl ether.
1910.1007 3,3'-Dichlorobenzidine (and its salts).
1910.1008 bis-Chloromethyl ether.
1910.1009 beta-Naphthylamine.
1910.1010 Benzidine.
1910.1011 4-Aminodiphenyl.
1910.1012 Ethyleneimine.
1910.1013 beta-Propiolactone.
1910.1014 2-Acetylaminofluorene.
1910.1015 4-Dimethylaminoazobenzene.
1910.1016 N-Nitrosodimethylamine.
1910.1017 Vinyl chloride.
1910.1018 Inorganic arsenic.
1910.1020 Access to employee exposure and medical records.
1910.1025 Lead.
1910.1027 Cadmium.
1910.1028 Benzene.
1910.1029 Coke oven emissions.
1910.1030 Bloodborne pathogens.
1910.1043 Cotton dust.
1910.1044 1,2-dibromo-3-chloropropane.
1910.1045 Acrylonitrile.
1910.1047 Ethylene oxide.
1910.1048 Formaldehyde.
1910.1050 Methylenedianiline.
1910.1051 1,3-Butadiene.
1910.1052 Methylene Chloride.
1910.1096 Ionizing radiation.
1910.1200 Hazard communication.
1910.1201 Retention of DOT markings, placards and labels.
1910.1450 Occupational exposure to hazardous chemicals in laboratories.

SUBJECT INDEX FOR 29 CFR 1910 - OCCUPATIONAL SAFETY AND HEALTH STANDARDS

Sample Permit Forms

Confined Space Entry Permit

Date and Time Issued: ______________ Date and Time Expires: ___________
Job site/Space I.D.: ______________ Job Supervisor:______________
Equipment to be worked on: __________ Work to be performed:____________
Stand-by personnel: ____

1. Atmospheric Checks: Time _______
   Oxygen _______%
   Explosive _______% L.F.L.
   Toxic _________PPM

2. Tester's signature: ____

3. Source isolation (No Entry):	N/A	Yes	No
Pumps or lines blinded,	( )	( )	( )
disconnected, or blocked	( )	( )	( )
	( )	( )	( )

4. Ventilation Modification:	N/A	Yes	No
Mechanical	( )	( )	( )
Natural Ventilation only	( )	( )	( )

5. Atmospheric check after isolation and Ventilation:
   Oxygen _________% > 19.5 %
   Explosive ________ % L.F.L < 10 %
   Toxic ___________ PPM < 10 PPM H(2)S
   Time ___________
   Testers signature: ____

6. Communication procedures:

____

7. Rescue procedures:

____

____

8. Entry, standby, and back up persons:	Yes	No
Successfully completed required training?	( )	( )
Is it current?	( )	( )

9. Equipment:	N/A	Yes	No
Direct reading gas monitor -tested	( )	( )	( )
Safety harnesses & lifelines for entry & standby persons	( )	( )	( )
Hoisting equipment	( )	( )	( )
Powered communications	( )	( )	( )
SCBA's for entry and standby persons	( )	( )	( )
Protective clothing	( )	( )	( )
All electric equipment listed Class I, Division I, Group D and Non-sparking tools	( )	( )	( )

10. Periodic atmospheric tests:

Oxygen	____%	Time ____	Oxygen	____%	Time ____
Oxygen	____%	Time ____	Oxygen	____%	Time ____
Explosive	____%	Time ____	Explosive	____%	Time ____
Explosive	____%	Time ____	Explosive	____%	Time ____
Toxic	____%	Time ____	Toxic	____%	Time ____
Toxic	____%	Time ____	Toxic	____%	Time ____

We have reviewed the work authorized by this permit and the information contained here-in. Written instructions and safety procedures have been received and are understood. Entry cannot be approved if any squares are marked in the "No" column. This permit is not valid unless all appropriate items are completed.

Permit Prepared By: (Supervisor)____
Approved By: (Unit Supervisor)____
Reviewed By (Cs Operations Personnel) :

____

(printed name) (signature)

This permit to be kept at job site. Return job site copy to Safety Office following job completion.

Copies: White Original (Safety Office)
Yellow (Unit Supervisor)
Hard(Job site)

Appendix A

ENTRY PERMIT

PERMIT VALID FOR 8 HOURS ONLY. ALL COPIES OF PERMIT WILL REMAIN AT JOB SITE UNTIL JOB IS COMPLETED

DATE: SITE LOCATION and DESCRIPTION

PURPOSE OF ENTRY

SUPERVISOR(S) in charge of crews Type of Crew Phone #

COMMUNICATION PROCEDURES

RESCUE PROCEDURES (PHONE NUMBERS AT BOTTOM)

ALL CAPS WITH ASTERISK * DENOTES MINIMUM REQUIREMENTS TO BE COMPLETED AND REVIEWED PRIOR TO ENTRY*

REQUIREMENTS COMPLETED	DATE	TIME
Lock Out/De-energize/Try-out	____	____
Line(s) Broken-Capped-Blanked	____	____
Purge-Flush and Vent	____	____
Ventilation	____	____
Secure Area (Post and Flag)	____	____
Breathing Apparatus	____	____
Resuscitator - Inhalator	____	____
Standby Safety Personnel	____	____
Full Body Harness w/"D" ring	____	____
Emergency Escape Retrieval Equip	____	____
Lifelines	____	____
Fire Extinguishers	____	____
Lighting (Explosive Proof)	____	____
Protective Clothing	____	____
Respirator(s) (Air Purifying)	____	____
Burning and Welding Permit	____	____

Note: Items that do not apply enter N/A in the blank.

**RECORD CONTINUOUS MONITORING RESULTS EVERY 2 HOURS

CONTINUOUS MONITORING**

Permissible ______________________________

Appendix A

TEST(S) TO BE TAKEN	Entry Level	
PERCENT OF OXYGEN	19.5% to 23.5%	______
LOWER FLAMMABLE LIMIT	Under 10%	______
CARBON MONOXIDE	+35 PPM	______
Aromatic Hydrocarbon	+ 1 PPM * 5PPM	______
Hydrogen Cyanide	(Skin) * 4PPM	______
Hydrogen Sulfide	+10 PPM *15PPM	______
Sulfur Dioxide	+ 2 PPM * 5PPM	______
Ammonia	*35PPM	______

* Short-term exposure limit: Employee can work in the area up to 15 minutes.
+ 8 hr. Time Weighted Avg.: Employee can work in area 8 hrs (longer with appropriate respiratory protection).

REMARKS:______

GAS TESTER NAME
INSTRUMENT(S)
MODEL ______
SERIAL &/OR & CHECK # USED ______
&/OR TYPE UNIT # ______

SAFETY STANDBY PERSON IS REQUIRED FOR ALL CONFINED SPACE WORK

SAFETY STANDBY CHECK # ______
CONFINED SPACE ENTRANT(S) CHECK # ______
CONFINED SPACE ENTRANT(S CHECK # ______

SUPERVISOR AUTHORIZING
- ALL CONDITIONS SATISFIED ______
DEPARTMENT/PHONE ______

AMBULANCE Phone # ______

FIRE Phone # ______

Safety 4901 ______

Gas Coordinator Phone #______

Appendix A

Sample LockOut/ TagOut Procedures

These sample procedures are taken from A guidelines.

General

The following simple lockout procedure is provided to assist employers in developing their procedures so they meet the requirements of this standard. When the energy isolating devices are not lockable, tagout may be used, provided the employer complies with the provisions of the standard which require additional training and more rigorous periodic inspections. When tagout is used and the energy isolating devices are lockable, the employer must provide full employee protection (see paragraph (c)(3)) and additional training and more rigorous periodic inspections are required. For more complex systems, more comprehensive procedures may need to be developed, documented, and utilized.

Lockout Procedure

Lockout Procedure for____

(Name of Company for single procedure or identification of equipment if multiple procedures are used).

Purpose

This procedure establishes the minimum requirements for the lockout of energy isolating devices whenever maintenance or servicing is done on machines or equipment. It shall be used to ensure that the machine or equipment is stopped, isolated from all potentially hazardous energy sources and locked out before employees perform any servicing or maintenance where the unexpected energization or start-up of the machine or equipment or release of stored energy could cause injury.

Compliance With This Program

All employees are required to comply with the restrictions and limitations imposed upon them during the use of lockout. The authorized employees are required to perform the lockout in accordance with this procedure. All employees, upon observing a machine or piece of equipment which is locked out to perform servicing or maintenance shall not attempt to start, energize, or use that machine or equipment.

Type of compliance enforcement to be taken for violation of the above.

Sequence of Lockout

(1) Notify all affected employees that servicing or maintenance is required on a machine or equipment and that the machine or equipment must be shut down and locked out to perform the servicing or maintenance.

____

Name(s)/Job Title(s) of affected employees and how to notify.

(2) The authorized employee shall refer to the company procedure to identify the type and magnitude of the energy that the machine or equipment utilizes, shall understand the hazards of the energy, and shall know the methods to control the energy.

____

Type(s) and magnitude(s) of energy, its hazards and the methods to control the energy.

(3) If the machine or equipment is operating, shut it down by the normal stopping procedure (depress the stop button, open switch, close valve, etc.).

____

Type(s) and location(s) of machine or equipment operating controls.

(4) De-activate the energy isolating device(s) so that the machine or equipment is isolated from the energy source(s).

____

Type(s) and location(s) of energy isolating devices.

(5) Lock out the energy isolating device(s) with assigned individual lock(s).

____

Individual lock numbers and personnel assignments

(6) Stored or residual energy (such as that in capacitors, springs, elevated machine members, rotating flywheels, hydraulic systems, and air, gas, steam, or water pressure, etc.) must be dissipated or restrained by methods such as grounding, repositioning, blocking, bleeding down, etc.

____

Type(s) of stored energy - methods to dissipate or restrain.

(7) Ensure that the equipment is disconnected from the energy source(s) by first checking that no personnel are exposed, then verify the isolation of the equipment by operating the push button or other normal operating control(s) or by testing to make certain the equipment will not operate.
Caution: Return operating control(s) to neutral or "off" position after verifying the isolation of the equipment.

____

Method of verifying the isolation of the equipment.

(8) The machine or equipment is now locked out.

"Restoring Equipment to Service." When the servicing or maintenance is completed and the machine or equipment is ready to return to normal operating condition, the following steps shall be taken.

(1) Check the machine or equipment and the immediate area around the machine to ensure that nonessential items have been removed and that the machine or equipment components are operationally intact.

(2) Check the work area to ensure that all employees have been safely positioned or removed from the area.

(3) Verify that the controls are in neutral.

(4) Remove the lockout devices and reenergize the machine or equipment. Note: The removal of some forms of blocking may require reenergization of the machine before safe removal.

(5) Notify affected employees that the servicing or maintenance is completed and the machine or equipment is ready for used.

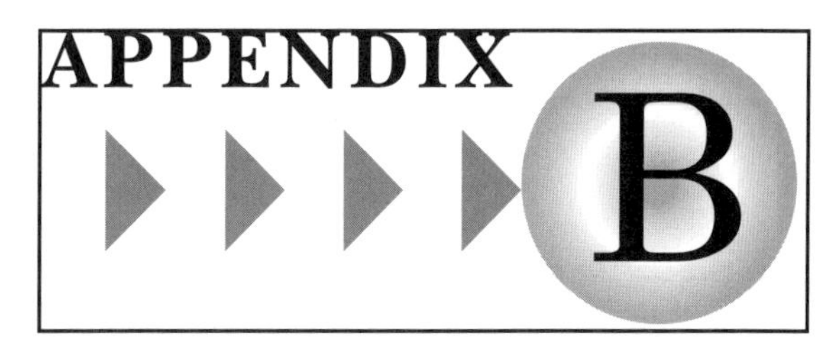

Sample FDA Compliance Checklist – FDA Supplied.

Here are some pointers for using the checklist:

1. Check the box beside each item to indicate a "situation under control," or one that needs further attention.
2. Use the space provided at the end of each topic section to note what you intend to do to collect an identified problem, and to note any compliance problems you face that are not addressed by the checklist. This checklist is a guide to be developed according to the needs of your operation.
3. Feel free to photocopy this checklist and to use it regularly during your inspections.

Employees

We'll begin the check with your employees. They are your most important resource.

OK		Needs Attention
☐	Are the employees well-trained in what they do? You can avoid many problems by making sure that your employees clearly understand their functions.	☐
☐	In handling food products, do your employees wear the proper hair covering and clean uniforms?	☐
☐	Are your employees wearing jewelry, bandages, or have any illnesses, infections or injuries (i.e., boils, cuts) which can contaminate foods?	☐
☐	Do your employees wash after each visit to the toilet? Do you have washing facilities available for your employees near their work stations, and do they use them when their hands become soiled or contaminated?	☐
☐	You must display "reminder" posters in your rest rooms for employees to wash their hands.	☐

OK		Needs Attention
☐	Do your employees maintain clean personal habits? They should keep their hands away from body surfaces, which are loaded with bacteria.	☐
☐	Is the traffic within your plant controlled to prevent contamination of the processing areas?	☐
☐	Have your employees been told the reasons why they should undertake the above precautions?	☐
☐	Other Employee practices that need attention:	☐

____

____

____

Plant/Grounds

OK		Needs Attention
☐	Is the area around your firm clear of weeds, grass and brush? This sort of foliage can be an effective cover for pests to infiltrate your firm.	☐
☐	Is there any standing water on your ground which also attracts pests?	☐
☐	Other outside Plant/Grounds conditions that I want to look into:	☐

Building/Facility

OK		Needs Attention
☐	Do windows and doors seal tightly to ward off pests and contaminants?	☐
☐	Do windows have fine mesh screens to keep out insects? Will a pencil pass under the door? That's all the space required for a rodent to enter.	☐

OK **Needs Attention**

☐ Have all holes and cracks been filled so as not to provide hiding places or entry points for pests? ☐

☐ Not only should your firm be free of vermin and pests—there shouldn't even be evidence of the presence of domestic animals such as cats and dogs. ☐

☐ Are rest rooms cleaned regularly? ☐

☐ Are the hand-washing facilities furnished with paper or air hand dryers and soap? ☐

The hand-washing facilities should be furnished with running water at a suitable temperature for washing hands.

☐ They should provide effective hand-washing and sanitizing preparations. ☐

☐ Does the roof leak? This can add to the problems of humidity, standing water and product contamination. ☐

☐ Are the overhead lights covered with shields to prevent contamination of products by broken glass in case the lamps burst? ☐

Other **Building/Facility** problems that need to be addressed:

____

____

____

Equipment

OK **Needs Attention**

☐ Is all equipment which comes in contact with food cleaned and sanitized as often as necessary to prevent contamination of the product? You should follow appropriate cleaning schedules for each piece of equipment. ☐

OK		Needs Attention
☐	Is the equipment designed, or otherwise suitable, for us in a food plant? For example, equipment for handling or processing foods cannot contain polychlorinated biphenyls (PCB's), which are very toxic (this does not apply to electrical transformers and condensers containing PCB's in sealed containers)	☐
☐	Is there a build-up of food or other static material on the equipment? This can serve as a breeding place for insects and bacteria.	☐
☐	Is there any build-up or seepage of cleaning solvents or lubricants on your equipment which can contaminate foods? All repairs on equipment should be of a permanent nature (e.g., no bobby pins in place of cotter pins), as temporary repair parts can break or rupture and get in the food product.	☐
☐	Is the equipment hard to disassemble for clean-up and inspection? The more difficult it is, the less inclined you or an employee will be to clean it.	☐
☐	Is there a lot of "dead space" in or around the machinery where food and other debris can collect as a nest for insects and bacteria?	☐
☐	Can the surface of the equipment be sanitized? Wood is one material that cannot.	☐
☐	Other Equipment cleaning and maintenance issues that should be covered:	☐

Housekeeping

OK		Needs Attention
☐	Are trash, debris, and clutter picked up so as not to provide hiding places for pests?	☐
☐	Do employees eat and smoke only in designated areas?	☐
☐	Is the food spilled or uneaten by employees cleaned up quickly so as not to attract pests or breed bacteria? Has old rodent excreta been cleaned up so you can spot any new activity?	☐

Appendix B

Additional Housekeeping duties that must be attended to:

Garbage

OK		Needs Attention
☐	Is garbage quickly removed and dumped in appropriate bins? It should not sit around your facilities to attract pests.	☐
☐	Is the garbage kept covered? An open garbage pile is an excellent breeding ground for insects and rodents.	☐
☐	Other Garbage-handling problems that should be explored:	☐

Plumbing

OK		Needs Attention
☐	Is the water used in your firm from an approved source (either municipal supply or tested private source)?	☐
☐	Have you made sure there are no hoses left dangling in sinks or on the ground? Loss of pressure can cause a back flow that will contaminate your water supply.	☐
☐	Do your facilities have back flow and vacuum breaker valves to prevent contaminate your water supply?	☐
☐	Avoid standing water around your firm.	☐
☐	Other Plumbing needs that require attention:	☐

Humidity

OK		Needs Attention
☐	Does your building have dripping condensation or leaky plumbing which can contaminate foods?	☐
☐	Are you keeping the humidity in your operation low? Molds, insects and bacteria thrive in damp climate.	☐
☐	Other problems to attend to regarding Humidity:	☐

Appendix B

Temperature

OK		Needs Attention
☐	Are storage areas intended for room temperature subject to extremes of temperature, either hot or cold? This can damage foods.	☐
☐	For refrigeration storage, coolers should be kept at or below 40 degrees F.	☐
☐	For freezer storage, the temperature should be kept at or below 0 degrees F.	☐
☐	Are you keeping a record of temperatures for all storage areas on a regular basis?	☐
☐	Are you keeping your facilities at the proper temperature range? Insects love high temperatures, and their activity will pick up as the temperature goes up.	☐
☐	Additional Temperature-related difficulties to explore:	☐

Incoming Raw Materials

OK		Needs Attention
☐	Have you checked to see that the compartment door seals on the truck are intact?	☐
☐	Is there a clean smell when the compartment doors are opened, or are there signs of contamination such as petroleum distillate, putrefaction, or other off-odors?	☐
☐	Is any refrigerated compartment set at the proper temperature?	☐
☐	Are boxes properly stacked and intact?	☐
☐	Is there evidence of activity by insects, rodents or birds?	☐
☐	Is there evidence of the misuse of pesticides such as DDT tracking powder, 1080, or insect sprays?	☐
☐	Additional problems that should be dealt with on The FDA publication *Inspecting Incoming Food Materials* will provide further information on conducting an inspection of incoming food materials.	☐

Appendix B

Incoming Raw Materials:

____

____

Storage of Raw Materials and Products

OK		Need Attention
☐	Is the storage area over-crowded? Such a condition prevents adequate inspection and clean-up and also increases the likelihood of damage to products during handling.	☐
☐	Are products stored on pallets and at least 18 inches away from the walls? It is important to leave space for inspection aisles so that rodent and insect activity can be seen more readily. You might consider painting a white line on the floor along the walls to indicate inspection aisles.	☐
☐	Other **Storage** problems that should be corrected:	☐

Rotation

OK		Need Attention
☐	Are products stored on a first-in, first-out basis to reduce the possibility of contamination through spoilage?	☐
☐	Are old products kept in front of the new to help in the rotation process?	☐
☐	Are all incoming products dated to ensure a proper rotation of stocks?	☐
☐	Are items overstocked? This increases the chances of spoilage and contamination.	☐
☐	When checking containers for contamination, are dusty, faded or discolored containers checked first? They are obviously the most suspect items.	☐
☐	Additional issues to address on the **Rotation** process:	☐

Quarantine

OK		Need Attention
☐	Are all products spoiled by damage, insects, rodents or other causes stored in a designated "Quarantine Area" to prevent their contact with safe products?	☐
☐	Are such quarantined items disposed of quickly to prevent the development of pest breeding places?	☐
☐	Are incoming materials inspected for damage or contamination so that they can be rejected?	☐
☐	Other problems to address in the Quarantine procedure:	☐

Pest Control

OK		Need Attention
☐	If you hire an outside pest control operator. you should:	☐
☐	Check regularly on what the pest control operator is doing. Don't accept what he's doing on faith.	☐
☐	Check to see what poisons he is using. Make sure the poisons do not contaminate foods.	☐
☐	Learn where and how many bait stations there are. They should be placed so as not to present any chance of food contamination. They should be checked regularly.	☐
☐	Check to see if fumigators are being used. Do they represent a hazard to employees or food safety?	☐
☐	If doing your own exterminating, you should:	☐
☐	Know there is no such thing as an all-purpose pesticide, especially where foods are concerned. Get qualified advice before using any poisons.	☐
☐	Make a map showing locations of all traps, bait stations, etc., and check them regularly.	☐
☐	Put money into building maintenance if that will help solve your pest problems. For instance, don't rely solely on rodenticides to control your pest problem and leave gaps in the doors for the rodents to enter. Make sure	☐

those gaps are sealed. Extermination is a poor second choice, and will cost you as much, or more, in the long run.

☐ Other **Pest Control** situations to explore: ☐

Storage and Handling of Hazardous Materials

OK **Need Attention**

☐ Are materials such as pesticides, herbicides, cleaning solvents, lubricants and boiler compounds accessible for use only by authorized employees? This will help prevent accidents such as food contamination and employee injuries due to ignorance and misuse. ☐

☐ Additional situations to consider regarding **Hazardous Material Storage:** ☐

Labeling

OK **Need Attention**

☐ Are all hazardous materials kept in bottles, or drums, or boxes that reflect their dangerous nature? ☐

☐ Even non-hazardous materials should be labeled correctly. Several babies died in a hospital because salt was mistakenly used for sugar in their formulae. ☐

☐ Make sure that any labels you market comply with the Food, Drug, and Cosmetic Act and Fair Packaging and Labeling Act. ☐

Other questions on Labeling that need to be considered:

☐ FDA does not have the authority to approve labels prior to marketing, but it does have jurisdiction once the label is in interstate commerce. FDA will take legal action if a product is not labeled in accordance with the law. FDA is willing to provide comments on your labeling prior to marketing, if you desire. ☐

Food Additives

OK **Need Attention**

☐ Make certain that the food additives you use are suitable and safe for the intended purposes. ☐

Other issues regarding **Food Additives** to be resolved:

Product Codes

OK		Need Attention
☐	Do you have an effective recall procedure set up?	☐
☐	Other considerations on Product Codes:	☐

By completing this brief inspection "patrol," you now have an idea of what the FDA investigator will generally look for when he visits your firm. This "short course" is far from complete, but it should provide a foundation to help you maintain a safe, quality food processing and storage operation.

Here are some last-minute hints to help you in your inspection and sanitation efforts:

1. As you inspect, use the checklist to make a record of the problems you encounter so you won't forget them. You can then make corrections based on the checklist.
2. Formulate inspection, clean-up and maintenance schedules and stick to them.
3. Define your employees' responsibilities; make sure each one understands his duties so that no essential details are ignored.
4. Be diligent in your sanitation efforts. The struggle to control pests, bacteria and the other problem areas is a fulltime effort.

You've just taken your first big step in the campaign for better food processing and storage. By reading this booklet, you've gained an awareness of the problems you might face, tactics for dealing with them, and knowledge that FDA is ready to help you with advice and further information on how you can deal with specific problems you encounter.

By taking preventive measures now, you can avoid potentially costly, mandated adjustments that might arise when the FDA investigator pays you a visit—and you can ensure that only quality, safe food products find their way to the consumers...a move we all want.

Sample Violations Selected From Regulatory Agency Files

The examples that follow are taken from the OSHA regulatory agency files, including the Federal Register. While there are numerous other cases from the other regulatory agencies, this appendix focuses on a sampling of the OSHA violations. Due to the fact that not all of the cases are finalized and some are still in the settlement phase, the names of the companies have been withheld. The focus of this appendix is not on the specific companies, but rather on the fines from non-compliance with the regulations. The examples included in this appendix illustrate the regulations referenced throughout the textbook. This appendix also highlights the interrelationship between the regulations and maintenance strategies.

General Infractions

The following section details companies that were found in violation of multiple regulations. The details are provided in each of the citations.

Multiple Infractions

In this next example, a company was also cited for across-the-board violations of OSHA policies. The agency made the following findings and proposed penalties of $33,475 based on the following information.

The following citation from OSHA is for a company at which OSHA alleged that the employer had twenty-eight serious violations, including;

- failure to inspect and maintain a freight elevator.
- failure to provide standard railings for open sided platforms and stairways.
- failure to properly guard open sided platforms, floor hole openings, and stairways.
- failure to provide a hearing conservation program.
- failure to properly transfer flammable liquids through self-closing valves and provide covers for flammable liquid containers.
- failure to provide lockout/tagout devices for machinery.
- failure to remove a defective forklift from operation.
- failure to properly inspect steel alloy chain slings.
- failure to properly guard machinery.
- failure to establish a written respiratory protection program.
- failure to develop and implement confined space entry procedures for employees.

In this case, the serious violations carried a total proposed penalty of $33,475. The company was also alleged to have seven other-than-serious violations, including:

- failure to maintain a fully charged fire extinguisher.
- failure to maintain the proper psi for compressed air used for cleaning purposes.
- failure to provide fixed wiring instead of extension cords for equipment.
- failure to assure the integrity of grounded conductors.

A serious violation is defined as a condition for which there is a substantial possibility that death or serious physical harm can result. An other-than-serious violation is a hazardous condition that would probably not cause death or serious physical harm, but would still have a direct and immediate relationship to the safety and health of employees.

Failure to Follow Up Violations

Some violations are cited when a company fails to take corrective action in response to a previously-cited violation, as the following example and article illustrate.

[The agency filed] a penalty of $45,000 against the firm for failing to abate a violation of OSHA standards.

The Federal Register documented the following information concerning the lack of corrective action.

> According to the OSHA area director, the action results from a follow-up investigation conducted from August 30 to November 17 in response to [the company's] failure to verify that it had corrected hazards previously cited by OSHA in June, 1999.
>
> OSHA alleges that the firm allowed employees to be overexposed to crystalline silica, and failed to conduct fit-testing of respirators."Inhaling airborne silica can lead to silicosis, a disabling, nonreversible–sometimes fatal–disease of the lungs," an Agency spokesperson said. "That's why OSHA is conducting a national special emphasis program on the hazard. The National Institute of Occupational Safety and Health estimates that two million U.S. workers annually are exposed to crystalline silica. We want to protect the employees to the best of our ability."
>
> A failure to abate is a notice of additional penalty issued against an employer who has failed to correct a violation that has become a final order of the Occupational Safety & Health Review Commission.OSHA can apply penalties on a daily basis for hazards an employer fails to correct.

The plant was fined $19,200 for one serious citation which included several hazards involving electrical equipment, problems with defective gaskets on flip seals and flip seals that were not closed on weather proof receptacles. Also cited were deficiencies in lockout procedures which ensure that hazardous energy is controlled during maintenance operations. In this case, personal lockout devices were not used by indi vidual workers involved in group maintenance details.

An additional penalty of $47,500 was assessed against the plant for one repeat violation involving unlabeled circuit breakers and exposed live parts in circuit breaker cabinets and for using flexible cords instead of fixed wiring at fixed motors and other equipment.

Whistleblower Provisions

OSHA enforces whistleblower protections that prohibit discharging or otherwise discriminating against any employee because he or she filed a complaint or reported unsafe working conditions under the Occupational Safety and Health Act, the Surface Transportation Assistance Act, the Asbestos Hazard Emergency Response Act, and other statutes.

An employee of a residential treatment facility for the mentally challenged, who alleged that he was illegally fired for reporting a hazardous workplace condition, was vindicated by the U.S. Labor Department. In addition to having all references to being fired removed from his personnel files, he also received $15,000 in back wages.

Work Hazards

The following examples illustrate specific citations and proposed penalties against two different companies for their alleged violation of regulations governing work hazards. In the first case,

Total Proposed Penalties: $68,000

One alleged Willful violation, with a proposed penalty of $56,000 for:

- failure to provide fall protection for employees constructing a masonry wall.

Four alleged Serious violations, with $12,000 in proposed penalties, for

- failure to provide a place of employment free from recognized hazards likely to cause death or serious physical harm to workers in that employees were exposed to the hazard of being caught between the rotating superstructure of an excavator and an earthen embankment;
- failure to train employees in the hazards posed by rotating equipment and clearances between equipment and solid objects;
- failure to develop a fall protection training program and to provide such training to exposed employees;

- failure to properly slope or shore earthen embankments against possible collapse;
- nonfunctioning backup alarm on a truck.

Total Proposed Penalties: $8,500

One alleged Willful violation, with a $7,000 proposed penalty, for:

- failure to provide fall protection for employees constructing a masonry wall.

One alleged Serious violation, with a proposed fine of $1,500, for:

- failure to develop a fall protection training program and provide such training to exposed employees.

Explosion Investigation and Related Violations

In this next example, a company was cited for not following regulations that would have protected the employees from the explosion. The agency made the following findings and proposed penalties of $$22,350 based on the following information.

> The inspection was triggered by a fire and explosion at the scrap yard during installation of new gas lines to supply oxygen and propane for welding purposes. Three employees were injured in the accident, one losing part of a leg, another part of a foot and, the third victim, a lower arm.
>
> Citations addressed numerous welding violations, including failing to guard against mixing of fuel gases and oxygen.
>
> Both companies (the site owner and the contractor) were also cited for allowing employees to work in an unprotected trench with vertical walls and for failing to provide them with training on the hazards associated with trench work.
>
> As a result of the welding safety deficiencies, three workers suffered serious permanent injuries, and several others were exposed to potential cave-in while working in an unsloped, unshored excavation. A good safety and health program would have prevented this terrible tragedy."
>
> Serious violations, like those cited against the two firms, result when there is substantial probability that death or serious physical harm could result and that the employer knew or should have known of the hazard.

Chemical Hazards

In this next example, a company was also cited for not following regulations that would have protected the employees from an explosion at the worksite. The agency made the following findings and proposed penalties of $66,150 based on the following information.

The inspection also found that:

- the presence and levels of lead-containing paint on tanks and pipes had not been identified and measured;
- the use of appropriate respiratory protection had not been ensured at all times.
- chemical containers had not been properly labeled.
- an oxygen and an acetylene cylinder were stored too close to one another.

Specifically, the citations and proposed penalties encompass:

One alleged Willful violation, accountin for $49,500 in proposed penalties, for:

- failure to implement a site-specific safety and health program to identify, evaluate and control safety and health hazards posed to employees decontaminating and removing storage tanks which contained hazardous sustances; failure to obtain information on the chemical and physical properties of hazardous substances expected at the tank farm prior to employees coming on-site.

Eight alleged Serious violations, with $16,650 in penalties proposed, for:

- a tank which formerly contained flammable naphtha was not thoroughlycleaned, ventilated or adequatelytestedforaflammableatmospherebeforeanemployeebegancuttingitscollarbolts with an acetylene torch
- failure to conduct air monitoring to identify respiratory hazards prior to employees entering tanks which formerly contained hazardous substances;
- failure to provide appropriate level of respiratory protection to employeesentering said tanks;
- air monitoring to determine exposure levels to hazardous substances not conducted prior to workers entering tanks ;
- air monitoring to determine employee exposure to methylene chloride not done when an employee decontaminated a tank which contained that hazardous chemical;
- a record of methylene chloride monitoring not made;
- failure to conduct lead exposure assessment and identify the presence of lead-containing paint on tanks and pipes prior to starting the job;
- respiratory protection not used by an employee cutting tanks and pipes on which lead containing paint was present.
- drums containing chemicals were not properly labeled.

One alleged Other than Serious violation, with no cash penalty proposed, for:

- an oxygen cylinder and an acetylene cylinder were stored less than 20 feet apart.

Repeat Violations

In this next example, a company was also cited for not following regulations that would have provided employees with safe exits, machine guarding protection and other safety hazards. The agency made the following findings and proposed penalties of $54,350 based on the following information.

Four alleged Repeat violations, accounting for $45,000 of the proposed penalties, for:

- obstructed exit access in that exterior fire doors, when opened, blocked exit access from upper floors;
- employees exposed to unguarded horizontal shafting on ten yarn twister machines;
- missing or broken stopping and starting devices, handles and pedals on ten yarn twister machines;
- an ungrounded power cord; two ungrounded fans.

Four alleged Serious violations, with $9,350 in penalties proposed, for:

- employees not informed in writing within 21 days after audiograms determined a standard threshold shift in their hearing had occurred;
- supervision for correct use of hearing protectors not ensured;
- unguarded rotating heads on yarn twister machines;
- unguarded projecting shaft ends on yarn twisters;
- flexible power cord used in place of fixed wiring; defective or damaged
- electrical equipment and cords not removed from service for repair and testing.

One alleged Other than Serious violation, with no proposed penalty, for:

- failure to adequately monitor for noise exposure.

False Statements

In this next example, a company was also cited for making false statements to the investigators and not providing proper fall protection. The agency made the following findings and proposed penalties of $300,000 based on the following information.

> Two company officials were sentenced today following guilty pleas to charges that they made false statements to investigators from the Occupational Safety and Health Administration during the investigation of the death of an employee.
>
> Officials of the company previously entered a guilty plea on behalf of the corporation to a charge that the firm willfully violated federal fall protection regulations and caused the death of the worker.
>
> [The defendants] were each sentenced to six months imprisonment, three years supervised probation and fined $2,000 by U.S. District Court Judge
>
> In addition to the sentences for the company officials, [the company] was fined $300,000 and placed on five years probation, the maximum allowed by law. Terms of that probation include the company must notify OSHA when opening any new jobsite and must conduct a job site safety analysis at each site. Additionally, the firm must adhere to stricter fall protection measures. OSHA standards call for fall protection at 25 feet for steel erection.

Floor Condition Violations

In this next example, a company was also cited for not providing proper scaffolding safety protection. The agency made the following findings and proposed penalties of $48,081 based on the following information.

> Specifically, the agency spokesperson noted, the company is being cited for the following alleged violations:

Safety inspections

Four alleged SERIOUS violations, including proposed penalties totaling $6,206 for:

- failing to maintain floors of workrooms in a dry condition and failing to maintain drainage where wet processes were used;
- failing to guard rotating machine parts;
- failing to enclose the inclined rotating shaft of a mixer unit; and
- failing to place guards on dryer gears.

Four alleged REPEAT violations, carrying proposed penalties totaling $40,000 for:

- allowing use of a forklift truck with a defective tire and failing to examine all forklift trucks daily before placing them in service;
- unguarded rotating shaft end, unguarded revolving couplings and unguarded horizontal shaft;
- unguarded pulley and unguarded drive belt; and, lastly,
- unguarded in-going nip points on calendar rolls.

Health inspection

One alleged SERIOUS violation including a proposed penalty of $1,875 for:

- failing to provide adequate and suitable facilities for quick drenching or flushing of the eyes and body in areas where employees were exposed to injurious corrosive materials.

Nine alleged OTHER-THAN-SERIOUS violations with no proposed penalties for:

- failing to document training given to employees in personal protective equipment use and maintenance;
- failing to remove from service a forklift truck which was producing high levels of carbon monoxide in its exhaust; five alleged violations relating to limiting potential employee occupational exposure to blood borne pathogens;
- failing to ensure that each container of hazardous chemicals in the workplace was properly labeled as to contents, and labeled as to the hazard presented by their contents.

Documentation Violations

In this next example, a company was also cited for not providing proper excavation safety protection. The agency made the following findings and proposed penalties of $155,000 based on the following information.

> As a result, the company is being cited for four alleged WILLFUL violations, carrying proposed penalties totaling $154,000, for:

- failing to maintain an equivalent OSHA Log at their facility which is as readable and comprehensible as the OSHA 200 form; for calendar years 1996, 1997, 1998 and 1999,
- failure to record 71 recordable injury and illness cases;
- failure to make available supplementary records of occupational illness and injuries for the month of December 1996; and
- failure to make readily available complete and accurate OSHA 200 logs for the 1996, 1997, 1998 and 1999 calendar years.

The company is also being cited for one alleged "other-than-serious" violation, including a $1,000 penalty, for failure to retain the signed and certified annual summary of occupational injuries and illnesses for the 1996 calendar year.

During the investigation, it was learned that employees were discouraged from reporting signs and symptoms of cumulative trauma disorders.

Powered Platform Violations

In this next example, a company was also cited for not providing proper power platform safety protection. The agency made the following findings and proposed penalties of $28,800 based on the following information.

The cited conditions included:

- lack of an arresting device that would function to stop the platform in the event of a load failure
- failure to post the platform's rated load capacity (the maximum load allowed)
- failure to prevent employees from riding the freight platform between floors
- failure to equip the platform with car gates
- failure to interlock the shaftway gates with the freight platform to prevent them from being opened when the platform was not at a landing
- the gates were inadequate to prevent falls in that they were not capable of withstanding 200 pounds of force.

The company was also cited for the following items that were unrelated to the freight platform: an ungrounded compressor fan, unguarded blades on another fan, no rails on the stairways between the third and fifth floors of the building and failure to maintain an injury and illness log.

Seven alleged Serious violations, with $26,400 in penalties proposed, for:

- failure to furnish a place of employment free from recognized hazards likely to cause death or serious physical harm in that employees were exposed to:
- the hazards of having their body parts caught between the moving platform and the shaftway due to the lack of car gates and employees riding the platform between floors;
- the hazard of a hoist failure while loading/unloading stock or riding the platform between floors due to the lack of an arresting device and the lack of a rated load capacity for the platform; and
- the hazards of falling down the shaftway when the platform was not in a loading zone, being struck by the platform while it entered or left a landing zone, and having body parts caught between the moving platform and the shaftway;
- hoistway gates not capable of withstanding 200 pounds of force;
- stairs from third to fifth floors were not equipped with railings;
- the compressor fan for a freezer was not guarded against contact;
- another compressor fan was not grounded.

One alleged Other than Serious violation, with a proposed penalty of $2,400, for:

- failure to maintain the OSHA 200 illness and injury log.

Electrical Safety Violations

In this next example, a company was also cited for not providing proper electrical safety protection. The agency made the following findings and proposed penalties of $42,000 based on the following information.

Specifically, the company is being cited for two alleged Willful violations, with $42,000 in proposed penalties, for:

- an employee working in an aerial lift was not protected from overhead electrical lines by de-energizing and grounding of the lines, guarding by insulation or some other equivalent means of protection;
- an employee working in an aerial lift was not properly trained in the nature of electrical hazards associated with overhead electrical lines and the correct procedures for dealing with such hazards.

Failure to Comply

In this next example, a manufacturer of steel doors and frames was also cited for not providing proper electrical safety protection. The agency made the following findings and proposed penalties of $530,500 based on the following information.

OSHA today issued to the company 14 failure-to-abate notices, which can carry penalties for every day a company does not take corrective action. Six of the 14 were deemed egregious and assessed penalties for 40 days, because OSHA alleges the company knowingly failed to abate:

- unguarded press brakes and unguarded points of operation on power presses (one employee lost a finger tip in an accident);
- excessive noise in the workplace that can cause deafness;
- lack of a lockout/tagout program to protect against sudden startups of machines while they were being serviced;
- no training of employees in the use of fire extinguishers; lack of vermin control in the workplace (severe accumulations of pigeon droppings); and
- unprotected spray painting.

Nineteen additional OSHA citations issued today to [the company] carry proposed penalties of $142,900. They allege two willful safety violations, 10 serious safety and health violations, five repeat violations, and two other-than-serious violations. Proposed penalties — including those for the notices of failure-to-abate — total $673,400.

Guards, Electrical Safety, and Confined Space Violations

In this next example, the company was cited for not providing proper electrical safety protection. The agency made the following findings and proposed penalties of $31,950 based on the following information.

> The serious safety hazards, resulting in penalties totaling $10,500, included lack of point of operation guards on slitters and saws, an unguarded belt and pulley on a tilt saw, ungrounded electrical equipment and electrical cords not connected to devices and fittings to prevent tension to joints or terminal screws. Hazards associated with welding cables — using them with damaged insulation and with a splice at less than 10 feet from their holder — also fell under the serious category.
>
> OSHA also observed protective covers and devices missing from electrical boxes. An additional $10,000 fine was imposed for this violation, designated as repeat because the company had been cited previously for exposed live parts in 1997.
>
> The agency's health inspection at the site resulted in citations for three serious items, including the company's failure to evaluate the worksite to determine if spaces should be classified as permit-required confined spaces and to post signs at the location of permit spaces. Also designated as serious were violations involving blocked exits and various hazards associated with abrasive wheels. The total penalty for the serious violations was $3,750.
>
> Two repeat health violations, with a total penalty of $7,700, were cited because of poor housekeeping at the plant and not training employees on hazards associated with chemicals used in the plant.

General Maintenance Procedures Violations

In this next example, the company was cited for not providing proper briefing of safety procedures during a plant shutdown. The agency made the following findings and proposed penalties of $30,000 based on the following information.

> OSHA inspectors found that the hydrogen was released into the atmosphere by maintenance workers who were overhauling the generator. The workers had not been informed during a morning briefing that the hydrogen used to cool the generator during normal operations had not been purged as scheduled. Hydrogen is typically purged soon after the equipment is turned off and before disassembly begins, usually by the second or third day of the overhaul process which, on April 8, was in its 13th day.
>
> As a result of the inspection, OSHA cited the company for four serious violations involving maintenance procedures and proposed penalties totaling $25,200. The violations covered deficiencies in several areas, including "job briefing," lockout procedures which ensure that hazardous energy is controlled during maintenance operations, and personal lockout devices to be attached and removed by individual workers involved in the maintenance detail.
>
> In addition to the citation items that were related to the explosion at the plant, OSHA also issued citations pertaining to a complaint inspection at plant. The complaint was filed after the explosion at the plant. As a result of this inspection, OSHA issued three serious and one other-than-serious violation with proposed penalties of $4,875. The serious violations addressed hazards covering housekeeping, exposed chains and sprockets and an inadequate emergency response plan to address chemical spills. The other-than-serious violation covered the labeling of water pipes as "nonpotable" water.

Results of Poor Maintenance and Housekeeping

In this next example, the company had experienced an explosion in the plant ventilation system. The following is a description of the incident and the subsequent investigation. The item that has the most impact on the maintenance organization is the causal factors mentioned in the conclusion.

> He [the agency investigator] noted that these three agencies proactively undertook a cooperative, joint investigation into the cause of the explosion and fire, which extensively damaged several buildings in [the company's] Foundry complex and seriously injured twelve employees. Three of the most severely burned employees subsequently died from their injuries.
>
> He noted that the investigation determined that an initiating fire event

in one of [the company's] Mold stations in the Mold Building was pulled into the exhaust ventilation system. The interior of the ductwork of that system was heavily loaded with deposits of phenol formaldehyde resin, an explosive organic dust. The ignition of this dust caused a turbulent fire and explosion(s) which traveled through the interior ductwork and in turn shook down explosive concentrations of combustible resin dust that had collected on surfaces throughout the Mold Building. When the fire exploded out from the ductwork, it ignited these airborne concentrations of combustible dust, causing a catastrophic dust explosion which lifted the building's roof and blew out its walls.

The report notes that, although it was not possible to conclusively determine the initiating event which caused the resultant dust explosion, a number of plausible scenarios were developed from the physical and testimonial evidence. Of these, the following two were determined to be the most probable:

- Dust Scenario: Heavy deposits of resin dust were found in the flexible exhaust ducts serving the ovens in the shell molding stations. The open ends of the ducts were placed adjacent to the ovens, at approximately head level, and in an area where employees must present themselves to deal with the ovens. Jarring of the duct readily dislodged the deposits of dust. In this scenario, jarring of the duct caused dust to fall down onto the oven and be ignited. The resulting fireball was then pulled back into the flexible duct where it started the turbulent fire and explosions in the exhaust ventilation system.
- Gas Scenario: The fuel trains to the ovens in the shell molding stations were found to be in very bad condition. The internal mechanisms of the valves controlling the flow of combustion air and natural gas to the ovens were found to be massively contaminated with resin and sand. The proper functioning of these valves was critical for providing air and gas to the ovens in the correct ratio to support combustion. Oven flameouts were a recurrent problem. The ovens were not provided with a flame-sensing device to prevent the flow of gas to the oven in the absence of main flame. Although a switch and thermocouple prevented the flow of gas to the oven in the absence of a pilot flame, the pilot flame was not able to light the burners. Thus, in the absence of a main flame, gas could continue to flow to the oven. In this scenario, gas was flowing to an oven that was not lit. The unburned gas collected in sufficient quantities to finally be ignited by the pilot or other ignition source, and the resulting fireball was pulled into the flexible duct where it started the turbulent fire and explosions in the exhaust ventilation system.

The report found that inadequate housekeeping, ventilation, maintenance practices and equipment were all causal factors for the initiating and catastrophic events.

General Maintenance and Storage Violations

In this next example, the company was cited for not providing proper guarding and storage of compressed gas cylinders. The agency made the following findings and proposed penalties of $175,000 based on the following information.

The company was cited for 16 alleged repeat violations carrying a penalty of $160,600, including deficiencies in guarding power presses, shafts, belts, pulleys, conveyors, and other machinery; electrical hazards; failure to maintain adequate aisle widths; storage of oxygen and acetylene cylinders together; and not conducting adequate inspections of power presses.

The alleged serious violations, carrying a total penalty of $14,000, for which the employer was cited included:

- failure to guard openings in the floor;
- failure to comply with requirements for mechanical power presses;
- failure to properly guard rotating shaft couplings;
- failure to equip fan motors with means to be disconnected;
- failure to provide training or fit-testing to employees using respirators.

The firm was also cited for having aisles restricted by parked vehicles, missing railings on stairways, and lock-out locks used for other purposes, three alleged other-than-serious violations carrying a total penalty of $400.

General Safety and Health Inspections

In this next example, the company was cited for not providing proper guarding of equipment, electrical safety and scaffolding violations. The agency made the following findings and proposed penalties of $190,400 based on the following information.

Safety inspection

- Thirty-nine alleged SERIOUS violations, including proposed penalties totaling $86,500, relating to machine guarding, fire extinguisher and electrical hazards and the use of defective powered industrial trucks; as well as maritime violations relating to unguarded manholes, decks and floor openings; blocked exits and fire extinguishers; chains not marked or inspected; loads suspended over workers' heads; and personal fall arrest systems not adequate.

- Eight alleged REPEAT violations, carrying proposed penalties totaling $85,400, for maritime hazards such as unsafe scaffolds or staging; missing mid-rails on scaffolds, staging or working platforms; defective ladders; lack of safety belts and lifelines; and housekeeping. [The company has been cited previously for similar violations.

• Nineteen alleged other-than-serious violations, carrying no proposed penalties, relating to egress and electrical hazards; damaged wire rope slings; improperly stored compressed gas cylinders; and unsecured floor hole covers.

Health inspection

• Six alleged SERIOUS violations, including proposed penalties of $18,500, relating to inadequate eyewash stations, hazard communication, personal protective equipment and ventilation hazards; and maritime violations relating to personal protective equipment and painting hazards.

• Seven alleged other-than-serious violations, with no proposed penalties, relating to hazard communication and personal protective equipment hazards.

Prior inspections

The most significant case dates from 1987 when a comprehensive inspection of the facilities resulted in proposed penalties totaling $4,200,000.

Recordkeeping Violations

In this next example, the company was cited for not maintaining proper OSHA logs, including proper employee audiograms. The agency made the following findings and proposed penalties of $68,000 based on the following information.

> The willful violation, carrying a proposed penalty of $50,000, concerned the company's failure to record injuries and illnesses on OSHA logs for 1996,1997, 1998, and 1999. These unrecorded cases included numerous instances of musculoskeletal disorders.
>
> Additional fines totaling $18,000 were assessed because the company did not abate a record keeping practice involving standard hearing threshold shifts. OSHA requires that employers record results of worker audiograms showing hearing losses of 25 decibels or greater.

Ergonomic Violations

In this next example, the company was cited for not providing proper ergonomic conditions in the workplace. The agency made the following findings and proposed penalties of $68,000 based on the following information.

> [The company representatives] have reached a joint settlement agreement with OSHA addressing the cited violations. According to the agreement, the company will pay the $68,000 fine, make ergonomic upgrades to certain equipment, conduct ergonomic analyses on jobs identified during the inspection and take appropriate action, and provide training for safety teams, managers, and employees.

Appendix C

General Violations Related to Preventive Maintenance Inspections

In this next example, the company was cited for numerous violations after a fire and explosion in the plant prompted a detailed inspection. The agency made the following findings and proposed penalties of $148,500 based on the following information.

> OSHA's inspection also identified an additional thirty-eight violations in other areas of the foundry which, though unrelated to the fire and explosion, would pose a hazard to employee safety or health if left uncorrected.
>
> These citations address instances of: a crane being operated with damaged running ropes and obstructed operator vision; unguarded or inadequately guarded moving parts on a variety of machines; lack of guardrails or equivalent fall protection at different locations; damaged fork trucks not removed from use; unapproved electrical equipment used in wet locations, ungrounded or unlabeled electrical equipment; noise monitoring not conducted when required; various deficiencies in the provision, use or fit-testing of respirators, deficiencies in the company's program for handling emergencies in permit-required confined spaces; and employees exposed to excessive levels of coal tar pitch volatiles, silica and a mixture of ammonia, phenol and formaldehyde where feasible engineering controls had not been implemented to reduce those exposure levels.
>
> The company also agrees to implement a comprehensive safety and health program by October 1, 1999, for all its operations. The program will include, but not be limited to:
>
> - Employing a fulltime safety and health manager.
> - Establishing a preventive maintenance program which includes a data base for checking occurrence, frequency and repair of all maintenance problems, including provisions to ensure proper operation of shell molding fuel trains and prevention of accumulation of explosive resin dust.
> - A written affirmation by the company's president of the fundamental importance of worker safety and health through the issuance of a clear policy on safe and healthful working conditions to all current employees and new hires.
> - Establishing clear safety and health goals for management, with clear responsibilities, goals and accountability, and overseen by senior management.
> - Committing sufficient resources and personnel to:
>   - conduct annual safety and health program reviews to improve the company's ability to identify and correct hazards;

- conduct quarterly worksite inspections to identify and eliminate specific hazards;
- provide for employee safety ands health training, primarily to address conditions and hazards cited in safety and health inspections.

In addition, in rebuilding its mold department, the company shall utilize state-of-the-art equipment and systems with particular attention to safety and health features. Any ventilation system will be designed and approved by a certified ventilation engineer and all shell molding stations and associated ventilation will conform to National Fire Protection Association guidelines. The company will also examine its other shell molding operations and equipment and systems replaced to conform to state-of-the-art no later than November 1, 1999.

Noise Exposure, Guarding, and Forklift Truck Violations

In this next example, the company was cited for numerous violations related to forklift safety. The agency made the following findings and proposed penalties of $42,500 based on the following information.

The alleged serious violations for which the employer was cited included:

- failure to have forklift truck operators wear seat belts;
- failure to provide fixed ladder with uniform rung spacing;
- failure to properly guard floor openings;
- failure to maintain exit passageways;
- failure to implement a training program for noise exposure;
- failure to provide personal protective equipment;
- failure to provide lockout-tagout procedures, and periodic review of a training program, to prevent the accidental start-up of machinery during repair or servicing;
- failure to maintain fire extinguishers and conduct fire extinguisher training;
- failure to provide adequate guarding of ingoing nip points, points of operation, rotating parts and other moving parts of machinery;
- failure to properly maintain forklift trucks;
- failure to maintain bench grinders;
- failure to reduce the pressure of compressed air used for cleaning to less than 30 psi;
- failure to cover exposed electrical equipment;
- failure to have electrical-safety training and work procedures.

Alleged other-than-serious violations for which the firm was cited included:

- not keeping aisles and passageways clear and in good repair,
- not displaying floor loading sign and exit signs, and
- not having a respiratory protection program.

Others included:

- using extension cords as permanent wiring,
- failure to label circuit breakers in electrical panels, and
- failure to identify the contents of containers of chemicals.

General Safety Violations

In this next example, the company was cited for not providing proper LOTO procedures, proper guarding or equipment, and proper inspections of mechanical presses. The agency made the following findings and proposed penalties of $67,050 based on the following information.

Among the alleged serious violations for which OSHA cited the company are:

- not posting appropriate signs indicating which doors are and are not exits.
- not establishing an adequate written energy control program
- not developing procedures to control potential hazardous energy
- not providing hardware for lockout/tagout operations
- not providing training to employees on how to control hazardous energy.
- not properly guarding and grounding machinery such as mechanical power presses, band saws, lathes, table routers and staple machines.
- not providing proper insulation for flexible electrical cables.
- not periodically inspecting mechanical power presses.

The alleged serious violations carry a total proposed penalty of $67,050.

Trenching and Construction Violations

The following citations are related to trenching and general construction violations.

Trench Violations

Employees who work in trenches are protected by their own set of specific guidelines. In one case, an inspection generated the following citations and proposed penalties against a company:

One alleged Willful violation, with a proposed penalty of $55,000, for:

- employees working within a trench that was not adequately sloped or shored against collapse.

One alleged Serious violation, with a proposed penalty of $2,800, for:

- a damaged synthetic web sling was not removed from service.

Construction Violations

In this next example, a company was also cited for not providing pro-per trench protection. The agency made the following findings and proposed penalties of $126,000 based on the following information.

> As site development and underground subcontractor, [the company] was responsible for cutting and re-routing an underground sewer. Company employees were preparing to perform the work when an OSHA compliance officer observed them in an unprotected trench nearly eight feet deep, four feet wide and 12 feet long. The OSHA inspector advised site supervisors to correct the hazards, but on two separate return visits during the day found that no action had been taken.

Even after being ordered by the site's general contractor to stop work until OSHA standards were met, [the company] continued to allow employees to work in the trench without adequate protection. To meet a deadline and complete the pipe replacement and tie-in work overnight, the trench was dug several feet deeper. By the time the OSHA inspector returned the next day, he found workers in an unsloped, unshored 14-foot-deep trench with nearly vertical walls.

As a result, OSHA fined the company $70,000 for one willful violation of trench safety standards which require that employees be protected from cave-in hazards while working in an excavation.

One repeat violation, with a penalty of $35,000, was also cited because excavated materials were left less than two feet from the top edge of the excavation causing greater likelihood of a cave-in because of the increased weight of soil at the edge of the trench.

The remaining $21,000 penalty was proposed for three serious safety violations. These included not providing a safe means of exiting the excavation, allowing employees in the trench to work underneath loads handled by a backhoe bucket, and allowing employees to work on the faces of slopes just above others working in an unprotected excavation.

Scaffolding Violations

In this next example, a company was also cited for not providing proper scaffolding safety protection. The agency made the following findings and proposed penalties of $150,800 based on the following information.

OSHA cited the following alleged scaffolding safety violations:

Four alleged WILLFUL violations, carrying total proposed penalties of $140,000, for:

- failing to ensure that a wooden scaffold system was erected in accordance with the design of a qualified person;
- failing to provide a safe means of access to each of the three working levels of a 20 foot high wood scaffold system;
- allowing employees to work at elevations up to 20 feet on scaffolding with damaged or weakened parts (scaffold planks and brackets were not able to support their own weight and four times the maximum intended load); and,
- failing to ensure that a guardrail system was installed along all open sides and ends of platforms on a wooden scaffold system.

Six alleged SERIOUS violations, including proposed penalties totaling $8,600, for:

- employees potentially exposed to dropped tools or materials not protected by protective head gear;

- employees exposed to falling nails not protected with protective eye wear;
- the 2"x6" boards used as scaffold planks were placed so their ends extended up to four feet beyond the end supports;
- a competent person did not inspect the scaffold as required;
- toeboards and screens were not provided to protect workers from falling materials; and,
- employees were exposed to falls while standing unprotected on a second floor balcony railing.

A health inspection was also conducted, which resulted in:

Two alleged SERIOUS violations, carrying proposed penalties totaling $2,200, for:

- failing to develop a written respirator program for employees required to mix silica containing products and failing to assess employee exposure to respirable quartz and particulates;
- failing to develop and implement a written hazard communication program and failing to conduct hazard communication training.

A willful violation is defined by OSHA as one committed with an intentional disregard of, or plain indifference to, the requirements of the Occupational Safety and Health Act and regulations.

A serious violation is defined as one in which there is substantial probability that death or serious physical harm could result, and the employer knew, or should have known, of the hazard.

Trench Violations

In this next example, a company was also cited for not providing proper excavation safety protection. The agency made the following findings and proposed penalties of $81,000 based on the following information.

OSHA cited the company for two alleged willful violations of OSHA standards for failure to properly protect employees from cave-ins while working in an excavation, and failure to ensure that employees wore protective helmets while working in areas where there is danger of head injury from impact or from falling objects. The alleged willful violations carry a total proposed penalty of $70,000.

The company was also cited for four alleged serious violations of OSHA standards:

- failure to maintain a construction industry safety and health program.
- failure to properly instruct employees in the recognition and avoidance of worksite hazards.
- failure to keep employees from working in an excavation in which water had accumulated.
- failure to store excavated soil the required distance from the edge of an excavation.

The alleged serious violations carry a total proposed penalty of $11,000.

Trench Violations

In this next example, a company was also cited for not providing proper trench safety protection. The agency made the following findings and proposed penalties of $72,000 based on the following information.

Two alleged Willful violations, with $56,000 in penalties proposed, for:

- failure to have a protective system in place to prevent cave-ins in a trench five feet or more in depth;
- a four-foot high pile of excavated spoils was placed at trench's edge, where it could fall back into the trench.

Five alleged Serious violations, with $16,000 in proposed penalties, for:

- no ladder or other safe means of exit from the trench was provided to workers;
- failure to protect workers by backfilling, covering or barricading a trench more than six feet in depth where the trench could not be readily seen by employees working near the trench;
- workers were not trained to recognize and avoid trenching hazards nor instructed in correct procedures to prevent trenching hazards;
- failure to have a competent person conduct daily inspections of the trench to identify and correct hazards;
- employees were not wearing hard hats on the jobsite.

Trenching Violations

In this next example, the company was cited for not providing proper trenching safety protection. The agency made the following findings and proposed penalties of $45,800 based on the following information.

Specifically, the citations and proposed penalties encompass:
One alleged Willful violation, with a proposed penalty of $28,000, for:

- employees exposed to serious injury while working in an 8-foot deep trench which was not shored nor sloped to prevent a collapse of its sidewalls.

Two alleged Repeat violations, with proposed penalties of $11,200, for:

- employees working in an 8-foot deep, 44-foot long trench were not provided with a ladder or other safe means of exit
- failure to have the trench inspected by a competent person who could spot and authorize correction of hazards

Three alleged Serious violations, with $6,600 in penalties proposed, for:

- a concrete block catch basin that was undermined during trenching operations was not supported to prevent its collapsing into the trench in which employees were working;
- undermined asphalt pavement was overhanging the trench in which employees were working;
- a trench box was used with missing components.

Trenching, Fall Protection, and Guardrail Violations

In this next example, the company was cited for not providing proper trenching safety protection. The agency made the following findings and proposed penalties of $40,500 based on the following information.

Specifically, the citations and proposed penalties encompass:

One alleged Willful violation, with a proposed penalty of $33,000, for:

- employees exposed to falls into a six-to-nine foot deep, 175-foot long excavation that was not fenced or barricaded;

Two alleged Repeat violations, with proposed penalties of $6,000, for:

- fourteen instances where employees were exposed to falls of more than six feet from numerous wall openings on the first and second floor that were not provided with top rails;
- employees exposed to falls of up to 30 feet through a floor opening which lacked a midrail;

One alleged Serious violation, with a proposed penalty of $1,500, for:

- ten instances of wire rope guardrails on the first and second floors and roof area that were inadequate in that they were insufficiently taut to prevent employee falls when pushed in a downward direction; wire rope guardrails on the first and second floors and the roof area were not flagged with highly visible material.

Construction Site Violations

In this next example, the company was cited for not providing proper training and enforcement concerning moving equipment regulations. The agency made the following findings and proposed penalties of $45,500 based on the following information.

Specifically, the citations and proposed penalties encompass:

One alleged Willful violation, with a proposed penalty of $38,500, for:

- failure to provide a place of employment free from recognized hazards in that employees were exposed to the hazards of falling off and being struck by moving equipment due to their being allowed to ride a moving backhoe.

Four alleged Serious violations, with $7,000 in penalties proposed, for:

- employees working in an excavation that lacked collapse protection; employees exposed to possible sidewall collapse while installing shoring; failure to support an undermined road way adjacent to an excavation containing employees; and employees working in an excavation that had not been inspected by a competent person capable of recognizing and correcting potential hazards.

One alleged Other than Serious violation, with no cash penalties proposed, for :

- a steel sling was not marked with required information including its lifting capacity.

Personal Protective Equipment

The following citations are primarily related to violations of the PPE regulations.

Multiple Hazards, including Personal Protective Equipment

In this next example, a company was also cited for across-the-board violations of OSHA policies. The agency made the following findings and proposed penalties of $49,000 based on the following information.

1. failure to provide a place of employment free from recognized hazards likely to cause death or serious physical harm in that an employee who entered the freezer was exposed to carbon dioxide gas in excess of the immediately dangerous to life and health (IDLH) concentration of 40,000 ppm;
2. failure to supply the employee with an appropriate respirator;
3. failure to ensure that a second employee was stationed outside the freezer whenever an employee entered it, in order to monitor, provide assistance and maintain communication with the first employee;
4. no danger tag was posted on the freezer;
5. lack of a written hazard communication program;
6. failure to address how outside contractor employees would be informed of onsite hazardous materials and conditions and necessary precautionary measures;
7. employees were not trained on how to recognize and protect themselves from carbon dioxide hazards.

In addition to the previous fines, OSHA proposed additional fines for the items discovered during the investigation.

[The company] faces $28,000 in proposed penalties for four alleged Serious violations for:

1. failure to provide a place of employment free from recognized hazards likely to cause death or serious physical harm in that an employee who entered the freezer was overexposed to carbon dioxide gas in excess of the immediately dangerous to life and health concentration of 40,000 ppm;
2. failure to supply the employee with an appropriate respirator;
3. failure to ensure that a second employee was stationed outside the freezer whenever an employee entered it, in order to monitor, provide assistance and maintain communication with the first employee;
4. employees were not trained to recognize and protect themselves against the hazards of carbon dioxide gas and vapor.

Silicosis Exposure and Personal Protective Equipment

In this next example, a company was cited for not following regulations that would protect employees from silica exposure. The agency made the following findings.

[The agency] proposed penalties of $114,500 against the firm, for three alleged willful and eight alleged serious violations of OSHA standards. The company has until December 29 to contest the citations.

The action results from an inspection of the facility conducted from June 7 through November 30, in accordance with OSHA's special emphasis program for silicosis exposure.

OSHA alleges that the company willfully failed to protect its employees from silica exposure by:

- not implementing an effective respiratory protection program.
- not providing employees with the necessary respiratory protection training.
- not implementing an effective hazard communication program and not providing employees with required training about the health hazards of silica.

The alleged willful violations carry a total proposed penalty of $105,000.

"Inhaling airborne silica can lead to silicosis, a disabling, nonreversible, and sometimes fatal lung disease", OSHA also cited the company for the following alleged serious violations:

- not developing or implementing a written confined space entry permit program and not preparing entry permits for employees required to enter confined spaces.
- not providing training for employees on confined space entry permits.
- not documenting or using proper energy control procedures.
- not providing required employee training in the use of energy control procedures.
- not properly protecting employees from silica to prevent overexposure, and not using proper engineering or effective respiratory protection controls to limit employee exposure to silica.
- not properly storing compressed gas cylinders away from highly combustible materials.
- not adequately labeling hazardous chemicals, and not maintaining accurate material safety data sheets.

The alleged serious violations carry a total proposed penalty of $9,500.

A willful violation is defined by OSHA as one committed with an intentional disregard for, or plain indifference to, the requirements of the OSHA act and regulations.

Asbestos Exposure Violations

In this next example, the company was cited for not providing proper safety procedures for employees handling asbestos. The agency made the following findings and proposed penalties of $123,200 based on the following information.

The alleged willful violations addressed the company's failure to comply with the OSHA permissible exposure limit (PEL) for airborne asbestos exposure, failure to provide adequate engineering controls to limit airborne asbestos exposures, and failure to prohibit dry sweeping of asbestos dust and debris.

The alleged serious violations related to OSHA requirements for machine guarding, the asbestos standard, and the hazard communication standard. The machine guarding deficiency involved lack of point of operation guard-

.or riveting machines.
.ne apparent violations of the asbestos standard included

no additional air monitoring when changes were made in the brake pad production process that may result in exposures above the TWA (Time Weighted Average) limit;

regulated area not established when airborne asbestos concentrations exceeded the 8 hour TWA limit;

no respirators provided to workers as required by the asbestos standard;

lack of appropriate work clothing to protect workers from asbestos;

workers were permitted to take contaminated work clothing out of the change room;

it was not ensured that workers exposed above the TWA limit showered at the end of the workshift;

training not provided as required by the asbestos standard;

asbestos dust and debris were not disposed in sealed impermeable bags or closed containers; and

a medical surveillance program was not provided to all workers who were or would be exposed to airborne concentrations of asbestos above the TWA limit.

The apparent violations of the hazard communication standard involved deficiencies in the material safety data sheet which the employer was providing for its workers and for downstream users of asbestos brake pads.

Respiratory Protection, Hazard Release, and Confined Space Violations

In this next example, the company was cited for numerous violations of OSHA regulations, including not providing PPE, improper confined space procedures, and improper machine guarding. The agency made the following findings and proposed penalties of $206,550 based on the following information.

Among the thirty-five alleged serious violations of OSHA standards for which the employer was cited are:

- failure to develop and implement an adequate emergency action plan.
- failure to require employees exposed to excessive noise to use adequate hearing protection.
- failure to develop and implement an adequate emergency response plan in the event of a

release of a hazardous substance.
- failure to develop and implement an adequate respiratory protection program.
- failure to properly evaluate respiratory hazards at the facility.
- failure to develop and implement an adequate hazard communication program.
- failure to provide an adequate process safety management program for anhydrous ammonia, a highly hazardous chemical, and failure to implement appropriate safety systems to control releases of anhydrous ammonia.
- failure to obtain proper confined space permits and to provide adequate training for employees working in permit-required confined spaces
- failure to develop and implement a lockout-tagout program to control hazardous energy
- failure to provide an educational fire extinguisher program.
- failure to properly guard machinery.
- failure to use proper electrical components and comply with proper electrical practices.

The alleged serious violations carry a total proposed penalty of $107,550. OSHA also cited the company for ten other-than-serious violations, including:
- failure to adequately maintain the required OSHA log of occupational injuries and illnesses.
- failure to provide an appropriate eyewash and shower where corrosive materials were being used.
- failure to perform a survey to determine the presence, location and quantity of asbestos-containing material at the facility.
- failure to determine which employees were exposed to blood borne pathogens, to maintain adequate medical and training records for those employees, and to provide appropriate training for such employees.
- failure to assure employee participation and training in process safety management.

PPE, Chemical Exposure, and Fire Protection Violations

In this next example, the company was cited for numerous OSHA violations, including not providing proper PPE, improper machine guarding, and improper storage of compressed gas cylinders. The agency made the following findings and proposed penalties of $132,500 based on the following information.

The alleged serious violations for which the employer was cited included:

- overexposing employees to methylene chloride and failing to have required monitoring, leak detection, respiratory protection and other protective equipment, spill clean-up procedures, employee information and training, and other methylene chloride protective measures;
- allowing employees to eat in an area where the level of methylene chloride exceed permis sible exposure limits;
- failure to provide fire extinguisher training;
- failure to have an emergency response plan;
- failure to maintain means of egress;
- failure to properly guard moving parts on machines;
- failure to properly store cylinders of compressed gas;
- failure to provide fire extinguisher training;

- having a fire exit that discharged into a locked yard.

The firm was also cited for having inadequate rest room facilities, failing to maintain OSHA injury and illness logs, and not informing employees in writing of health monitoring results, three alleged other-than-serious violations carrying a total proposed penalty of $4,000.

Fall Protection

The following citations are related to violations of the fall protection standards.

Tower Incident

In this next example, a company was cited for not following regulations that would have protected the employees from the exposure to fall hazards. The agency made the following findings and proposed penalties of $18,400 based on the following information.

The worker fell to his death from the top of a 600 foot 911 tower near the building, which he had free-climbed to troubleshoot its strobe lights.

"Building and maintaining communication towers is a high growth industry," said an Agency spokesperson. According to the U.S. Labor Department's Bureau of Labor Statistics, 93 fatalities associated with this type of activity were reported from 1992 to 1997.

OSHA's safety inspection in response to the fatal accident revealed three serious violations. Among the hazards cited were:

a cage or ladder safety device was not used when employees climbed over 20 feet on the tower;

employees did not use a safety harness and tie off with a lanyard when free-climbing the tower, and

employees used tools that were not insulated or designed for live electrical work.

"We encourage companies to create an effective safety and health program," said an Agency spokesperson "If these companies can find and fix hazards prior to an OSHA inspection, they can save expensive penalties and, more importantly, they can save lives."

An Agency spokesperson added, "Taking proactive measures on safety and health issues can prevent this type of accident, reduce workers' compensation costs, improve employee morale, and ultimately increase company profits."

Fall Hazards

In this next example, a company was also cited for not following regulations that would have protected the employees from the exposure to fall hazards. The agency made the following findings and proposed penalties of $39,900 based on the following information.

OSHA's inspection found employees of both contractors exposed to a variety of fall hazards, in particular, potential falls of up to 50 feet through unguarded exterior wall openings on the second through fifth floors of the building.

Workers were also exposed to additional hazards involving the erection and use of scaffolding, including:

- employees erecting scaffolding without fall protection;
- an employee working in an aerial lift without fall protection;
- employees accessing the building's third floor by jumping from the scissors lift through a wall opening
- improperly installed scaffolding support posts
- employees not adequately trained nor knowledgeable about scaffold erection;
- scaffold erection not supervised by a competent person.

Nine alleged Serious violations, accounting for $17,100 of the proposed penalties, for:

- scaffold erection was not supervised by a competent person;
- employees accessed the third floor by jumping from an aerial lift through a wall opening;
- an employee operated an aerial lift without fall protection;
- scaffolding support posts were not installed in a level and sound manner;
- scaffold support posts were not installed plumb
- an unguarded wall opening;
- an unguarded floor hole and an unguarded stairwell;
- stairwell missing a handrail;
- a stairwell lacking guardrails.

Two alleged Repeat violations, accounting for $16,800 of the penalties proposed, for:

- employees erecting scaffolding without the use of fall protection;
- employees climbing cross braces to exit tubular a welded frame scaffold.

The company had previously been cited by OSHA for substantially similar violations in citations issued April 6, 1998, following an inspection at a Portsmouth, NH, worksite].

One alleged Other than Serious violation, with no proposed penalty, for:

- inadequately supported scaffolding midrails and toprails.

Four alleged Serious violations for:

- an unguarded wall opening;
- an unguarded floor hole and an unguarded stairwell;
- stairwell missing a handrail;
- a stairwell lacking guardrails.

Three alleged Other than Serious violation, with no penalties proposed, for: inadequate training and exposure assessment for Class IV asbestos work.

Bridge Fall Protection Violations

In this next example, a company was also cited for not following regulations that would have protected the employees from fall hazards at a bridge worksite. The agency made the following findings and proposed penalties of $22,800 based on the following information.

The alleged serious violations for which the employer was cited included:

- failure to provide ladders, stairing, ramps or other safe means of access between work levels with breaks in elevation greater than 19 inches;
- failure to provide fall protection on work surfaces higher than six feet;
- failure to provide anchorage points capable of supporting 5,000 pounds on fall-arrest systems;
- failure to limit free-fall distance on personal fall-arrest systems to six feet;
- failure to ensure that walking and working surfaces had the strength and structural integrity to support workers safely;
- failure to adhere to manufacturer's specifications on manlifts.

Fall Protection

In this next example, a company was also cited for not providing proper fall protection. The agency made the following findings and proposed penalties of $125,100 based on the following information.

As the result of the investigation, the firm was cited for the following alleged repeat violations:

- failure to provide guardrail system or personal fall arrest system;
- failure to provide adequate cross bracing on scaffolding;
- failure to provide an access ladder to reach the working level;
- failure to fully plank the working level;
- failure to provide protective headgear;
- failure to train employees in the recognition and avoidance of construction worksite hazards and the applicable OSHA standards.

The alleged repeat violations carry a total proposed penalty of $123,000.

The firm was also cited for not providing falling-object protection on the working level of the scaffold while other employees were on the level below without protective headgear, an alleged serious violation carrying a proposed penalty of $2,100.

Fall Hazards

In this next example, a company was also cited for not providing proper scaffolding safety protection. The agency made the following findings and proposed penalties of $69,400 based on the following information.

Specifically, the citations and proposed penalties encompass:

Two alleged Willful violations, accounting for $56,000 of the proposed penalties, for:

- employees exposed to serious fall hazards in excess of 35 feet while performing roofing work where a fall protection system was not in use
- employees exposed to serious fall hazards in excess of 35 feet while working within 3 feet of roof's edge during hoisting operations where standard guardrails were not provided and a personal fall arrest system was neither provided nor used.

Five alleged Repeat violations, with $8,400 in proposed penalties, for:

- roofing workers were not trained to recognize the unsafe working condi tions applicable to their work environment;
- employees were exposed to possible serious head injury while not wearing protective helmets;
- defective nylon slings used to hoist materials from ground to roof were not removed from service;
- the hydraulic crane was not inspected by a competent person (one with both the knowledge and authority to identify and correct hazards) before each use;
- annual inspection of the hydraulic crane had not been conducted.

Five alleged Serious violations, with $5,000 in penalties proposed, for:

- a written hazard communication program was not implemented for all hazardous materials such as asphalt and fiberboard, in use on the worksite;
- employees exposed to possible serious electric shock hazards while using ungrounded power tools;
- missing ground pin on an extension cord used to power electric equipment;
- employees exposed to serious injury while using a portable access ladder that did not extend three feet above the top of the access hatch;
- an 8-foot portable access ladder was not used according to manufacturer's recommendation.

The serious citation, and penalties of $19,000, are for the lack of ten inch high curbs or similar barriers at work areas on Pier #5, Barbers Point Harbor, where forklifts and other vehicles are used to load and unload ships along the 500-foot-long pier; and for failure to develop and disseminate an adequate emergency action plan for the work site for emergency response should an accident occur.

The willful citation against [the company] includes penalties of $70,000 for failure to provide curbing, or some other barrier along the water side edge of work areas where vehicles are used.

Last June, the company was cited for willful and serious violations and fined $135,000 following an OSHA investigation into an accident that seriously injured an employee who fell more than 30 feet from atop a stack of shipping containers. The employee, and another who also fell but was less seriously hurt, were not provided with required safety equipment to prevent falls.

Fall and Ladder Protection

In this next example, a company was also cited for not providing proper scaffolding safety protection. The agency made the following findings and proposed penalties of $179,500 based on the following information.

The firm was cited for eight alleged repeat violations carrying a total proposed penalty of $150,000, for not providing employees with fall protection, not ensuring that floor holes were covered, and not properly locating job ladders.

A repeat violation is one for which an employer has been previously cited for the same or a substantially similar condition and the citation has become a final order of the Occupational Safety and Health Review Commission.

The alleged serious violations for which the employer was cited included not providing eye protection, not providing hard hats, not providing fall-hazard recognition training, and not providing ladder-hazard recognition training, with a total proposed penalty of $29,500.

The firm was also cited for not providing enough toilets for the workers, an alleged other-than-serious violation.

Scaffolding and Fall Protection Violations

In this next example, the company was cited for not providing proper scaffolding and fall protection. The agency made the following findings and proposed penalties of $59,000 based on the following information.

> Consequently, [the company] is being cited for one alleged WILLFUL safety violation, carrying a proposed penalty of $56,000, for: failing to fully plank the work platform on a tubular welded frame scaffolding for the complete width of the platform; exposing employees to falls of 40 feet by not providing them with adequate fall protection on said scaffolding; and failing to provide employees working in a hoist area with a personal fall arrest system or guardrail system.
>
> The company is also being cited for one alleged SERIOUS violation, including a proposed penalty of $3,000, for failing to provide employees with an access ladder or equivalent safe access to the scaffolding.

Lock Out – Tag Out

In the following section, the citations deal with the violation of the Lock Out-Tag Out regulations.

Lock Out Tag Out

In this next example, a company was also cited for not providing proper Lock Out – Tag Out protection. The agency made the following findings and proposed penalties of $152,750 based on the following information.

The alleged serious violations for which the employer was cited included:

- failure to have a lockout-tagout program and training to prevent the accidental start-up of machines while they are being serviced or repaired;
- failure to provide personal protective equipment;
- failure to provide properly maintained, unobstructed, unlocked fire exits;
- failure to keep flammable liquids in covered containers when not in use, and to assure that an inside storage room for flammable liquids received a complete change of air at least six times per hour;
- failure to mark electrical circuits, correctly ground circuits and equipment, cover openings in fittings and boxes, and provide strain relief on extension cords;
- failure to have a written hazard communication program, to properly label containers of hazardous chemicals as to their contents, and provide an appropriate hazard warning on each container;
- failure to reduce below 30 p.s.i. the pressure of compressed air nozzles used for cleaning;
- failure to properly guard electrical saws, mechanical power presses, wood shapers, rotating parts on machines;
- failure to have a program of regular, periodic inspections of power presses;
- failure to train employees in the use of fire extinguishers.

The serious violations carry a total proposed penalty of $46,750.

Alleged other-than-serious violations included failure to maintain required logs of employee injuries and illnesses, keep the workplace clean and orderly, provide proper insulation on electrical connections, and properly mark exit signs.

Confined Space, Lock Out Tag Out Violations

In this next example, a company was also cited for not providing proper Lock Out – Tag Out protection. The agency made the following findings and proposed penalties of $53,750 based on the following information.

> The plant was targeted for inspection after an explosion in the plant's hardwood silo injured two employees.
>
> During the inspection, OSHA determined that lightning had caused the accident, but other safety hazards unrelated to the explosion resulted in citations for one willful violation with a proposed penalty of $41,250 and four serious violations with a proposed penalty of $12,500.
>
> The plant was cited for hazards associated with confined space, machine guarding and "lockout tagout," a procedure that cuts off energy sources so that machinery remains inoperative during servicing.
>
> "This company allowed employees to work inside the silo while the augers were in operation, a practice that could cause fatal injury," said [an OSHA

Spokesperson] "The augers, which were not de-energized, were crossed over by workers when entering and exiting the pine silo. In addition, while working inside the silo, workers were exposed to the moving augers."

LOTO, Guarding, and General Safety Violations

In this next example, the company was cited for not providing proper guarding, lockout/tagout, and proper disciplinary procedures for safety violators. The agency made the following findings and proposed penalties of $820,500 based on the following information.

OSHA inspected the plant as part of its ongoing efforts to examine workplaces with the highest injury and illness rates. As a result of this inspection, OSHA issued willful and serious citations for violations of:

- its machine guarding,
- lockout/tagout,
- mechanical power press,
- confined spaces and electrical standards.

In addition, the company agreed to review and revise its written safety and health programs to ensure that all employees and temporary workers are competent to recognize basic safety hazards and those that are unique to the industry.

The company also agreed to:

- train temporary, newly-hired and newly reassigned employees on specific work operations, machine guarding,
- lockout/tagout safety and

disciplinary rules to be applied in event of failure to observe work safety rules. The training will emphasize that no work is to be performed on machinery with removed, broken or disabled safety devices.

LOTO, Confined Space and Guarding Violations

In this next example, the company was cited for not providing proper LOTO procedures, confined space policies, and proper machine guarding. The agency made the following findings and proposed penalties of $151,000 based on the following information.

The serious citations are proposed for failing to provide fall protection in a number of cases, failure to provide personal protective equipment, machine guarding hazards, electrical lockout/tagout hazards, confined space entry hazards and for failing to keep potentially hazardous surfaces clean and dry.

The willful citations are proposed for:

- failing to develop and implement the means, procedures, and practices necessary for safe

confined space entry operations; and
- failing to prepare entry permits for permit-required confined spaces (PRCS) prior to authorizing entry.
- A repeat violation is proposed for failing to keep the chassis and wheel wells of fork trucks in a clean condition to prevent fires.

The other-than-serious citations are proposed for:
- not keeping concrete walkways free from holes;
- failing to secure a compressed gas cylinder against falling;
- failing to provide spill containment for an oil storage area;
- failing to empty buckets containing rainwater on a frequent basis;
- failing to install a sufficient number of emergency eye wash stations in a battery recharging room;
- failing to ensure that a portable fire extinguisher was readily accessible; and
- flexible cords were wrapped around the structural steel members of the building.

LOTO, Fall Protection, and Training Violations

In this next example, the company was cited for not providing proper LOTO policies and the related training. The agency made the following findings and proposed penalties of $245,000 based on the following information.

[The Company] received three willful violations, with a proposed penalty of $210,000 and seven serious violations, with a proposed penalty of $35,000.

The willful violations were due to the company's failure to:
- conduct annual reviews of lockout procedures, which are necessary to prevent the inadvertent start of machinery during maintenance operations;
- failure to conduct retraining on employees when the employer observed deviation from lockout procedures and
- failure to provide a walkway or fall protection for employees required to perform machine maintenance at an elevated work location.

The serious violations resulted from deficiencies in the company's lockout program, including:
- the absence of program enforcement procedures;
- inadequate employee training in Lockout/Tagout; and
- employee use of improper lockout procedures.

LOTO, Guarding, and Confined Space Entry Violations

In this next example, the company was cited for numerous general safety violations. The agency made the following findings and proposed penalties of $350,000 based on the following information.

[The Company] has agreed to pay a $350,000 penalty and to hire a consul-

tant to monitor its progress in correcting health and safety hazards,
As a result of OSHA's inspection, the company received citations for alleged deficiencies in programs and procedures relating to locking and tagging of machinery, confined space entry, and machinery guarding.

Additional violations involved hazards that could lead to slips and falls, or result in workers being struck by equipment or materials.

LOTO, Noise Abatement, and Emergency Response Violations

In this next example, the company was cited for not providing proper LOTO policies, noise exposure, and emergency response policies. The agency made the following findings and proposed penalties of $237,000 based on the following information.

OSHA's inspection resulted in citations against the company for 31 serious violations with a combined penalty of $127,000. The serious items included deficiencies associated with "lockout" safety procedures, noise exposure, use of respirators, walking/working surfaces, management of the anhydrous ammonia process, emergency response, and electrical and fire hazards.

Two willful violations found during the inspection drew additional penalties of $110,000. The willful violations included employee exposure to high levels of noise without hearing protection and inadequate training in the energy control, or "lockout," program.

Overhead Crane and Hoisting Violations

The following section contains citations related to overhead cranes and hoisting violations.

Crane and Hoist Inspection Violations

In this next example, a company was also cited for not providing proper overhead crane inspection and safety standards. The agency made the following findings and proposed penalties of $108,829 based on the following information.

The alleged serious violations for which the employer was cited included failure to:

- inspect and maintain overhead hoists, below-the hook lifting devices, and bridge crane systems;
- implement a crane preventive maintenance program;
- guard floor openings, floor holes, and ladderway openings;
- provide adequate railings for stairways;
- keep ladders In good repair and assure their proper use;
- maintain clear exit ways;

- keep aisles clear, dry, and in good repair;
- develop machine-specific lockout-tagout procedures to prevent the accidental start-up of machines during servicing or repair, train employees in the procedures, and require their use;
- properly secure and store compressed gas cylinders;
- maintain fire extinguishers and assure their accessability;
- protect sprinkler system standpipes, and assure that standpipe hoses and reels are accessible and serviceable;
- maintain inspection reports for cranes, ropes, and hooks;
- inspect chain and nylon slings and remove defective ones from service;
- adequately guard belts, pulleys, sprockets, and other machinery;
- provide and properly adjust work rests and tongue guards on grinders;
- guard live electrical parts;
- remove damaged electrical equipment from service;
- properly label circuit breakers and keep electrical switches accessible;
- provide fall protection;
- provide annual training, proper supervision, and annual hearing tests to employees exposed to high noise levels;
- properly dispense flammable liquids and take adequate precaution against ignition;
- require the use of goggles, hand protection, and other personal protective equipment;
- adequately label and test confined spaces and train personnel who work in or around them;
- label containers of hazardous materials;
- provide emergency eyewash stations.

The firm was also cited for alleged other-than-serious violations including failure to post floor load ratings, provide safe access to equipment, and properly label exits, and for using extension cords as permanent wiring, using electrical cords with splices and loose fittings, and using electrical outlets with reversed polarity.

Hoisting, Fall Protection and General Safety Violations

In this next example, the company was cited for not providing proper hoisting, fall protection and general safety protection. The agency made the following findings and proposed penalties of $38,700 based on the following information.

The alleged serious violations for which the employer was cited included:

- failure to keep hoists and other lifting devices in safe condition;
- failure to keep aisles and passageways clear and in good repair;
- failure to provide required fall protection on platforms;
- failure to have procedures for the lock-out or tag-out of machinery to prevent accidental start-up during servicing or repair;
- failure to maintain fire extinguishers and provide ready access to them;

- failure to provide emergency eyewash stations;
- failure to maintain powered industrial trucks;
- failure to post no-smoking signs at flammable storage area;
- failure to properly ground flammable solvent containers when dispensing fluid;
- failure to properly guard rotating shafts, belts, and other moving parts on machinery;
- failure to have cover plates on electrical panel boards and switch boxes.

The firm was also cited for alleged other-than-serious violations including failure to keep work floors clear and in a dry condition, to maintain flexible electric cords, post floor ratings, and label circuit breakers in electrical panels.

Fall Protection and Crane Violations

In this next example, the company was cited for not providing proper fall protection and permitting improper use of a crane as it related to lifting employees. The agency made the following findings and proposed penalties of $233,800 based on the following information.

> While on-site, the OSHA inspector observed employees working at heights ranging from 11 to 30 feet above the water without fall protection. The compliance officer also noted that a makeshift personnel platform used to raise workers from the ground to the bridge lacked guardrails and other required provisions needed to protect them from falls. In addition, the crane being used to lift the workers was not equipped with an anti-two-block device which prevents lifted items, like the personnel platform, from being raised so high that they hit the boom of the crane.
>
> These hazards resulted in three willful citations with penalties totaling $189,000.
>
> Eight serious violations of OSHA standards account for the remaining fines of $44,800. These addressed additional hazards associated with the crane and the fabricated personnel platform.

Crane and Hoisting Violations

In this next example, the company was cited for improper hoisting practices and other crane regulation violations. The agency made the following findings and proposed penalties of $223,000 based on the following information.

> OSHA cited the company for three willful violations with penalties totaling $195,000, for allowing employees to ride on and work under hoist-suspended loads. The company was also cited for failing to assure that loads were safely rigged before being hoisted so that employees could operate the hoists from a remote or other safe location.
>
> Additional penalties totaling $28,000 resulted from four serious related

violations.
"We categorized these citations as willful because the employer knew there was a hazard and took no action to protect workers," said [the OSHA spokesperson] "Even though the company's employee handbook explicitly prohibited it, our investigation revealed that employees routinely rode on and worked under loads suspended from hoists. In this case, the manner in which the hull plate was rigged required employees to stand on the plate to operate the hoists. In addition, the plate could not be fitted to the hull of the ship without going under the suspended load. With the hoists overloaded, this was a prescription for disaster."

[The OSHA spokesperson] added, "Supervisors, as well as rank and file employees, rode on and worked under hoisted loads routinely even though they were aware of OSHA's standards and the company's written prohibitions."

Noise Abatement, Hazard Assessment, and Crane Inspection Violations

In this next example, the company was cited for not providing proper hearing protection and necessary OSHA documentation. The agency made the following findings and proposed penalties of $88,000 based on the following information.

Specifically, the citations and $88,000 in proposed penalties encompass:
One alleged Willful violation, with a proposed penalty of $50,000, for:

- employees were exposed to excess noise levels, and a continuing effec tive hearing conservation program was not provided and all affected
- employees were not provided with annual hearing conservation training;
- employees using hearing protection were not refitted and retrained when audiometric testing showed a standard threshold shift indicating a permanent hearing loss; five instances where such threshold shifts were not recorded in the OSHA injury and illness log; and twenty instances where employees were not informed in writing in a timely manner when a review of their annual audiograms identified threshold shifts.

Fourteen alleged Serious violations, accounting for $29,500 in proposed penalties, for:

- two instances where employees were exposed to excess amounts of respirable crystalline silica and the employer did not institute feasible engineering controls to reduce the exposure level nor provide appropriate respiratory protection;
- all medical response technicians with occupational exposure to blood were not provided annual bloodborne pathogen training nor offered the Hepatitis B vaccine;
- an instance where an employee did not wear fire retardant clothing where required; the employer's written assessment of hazards in the workplace did not address a process

change which required the use of eye protection;
- failure to ensure that all components of an electrical pump used to transfer a flammable liquid were approved for use in a hazardous location;
- employees exposed to the hazard of being struck by improperly installed or maintained hoist components and/or loads;
- frequent inspections of cranes for defects were not conducted and crane operators did not receive sufficiently detailed training to identify excessive wear on cranes' running ropes;
- fall hazards posed by an unguarded floor hole, an inadequately guarded open-sided mezzanine and missing rails on stairways;
- improperly constructed wire rope slings; torn and damaged nylon sling in use;
  numerous instances of unguarded or inadequately safeguarded moving machine parts, unguarded or inadequately guarded points of operation on machinery and unguarded drive belts;
- several electrical safety hazards, including improperly installed or used electrical equipment; electrical conduits not protected against damage; uncovered junction boxes; ungrounded electrical equipment; power cords lacking grounding conductors; split and damaged insulation on power cords; extension cords used in lieu of permanent wiring; spliced extension cords; flexible cords pulled from their fittings.

One alleged Repeat violation, with a $7,500 penalty proposed, for:

- emergency eyewashes were not available for immediate use where employees worked with acids

The company was also cited for eight alleged Other than Serious hazards, with no cash penalties proposed, for:

- a blocked exit door;
- oxygen cylinder coated with dirt and wax;
- improperly constructed storage room for a flammable liquid;
- inadequate guard on an abrasive wheel;
- an improperly fitted compressed air nozzle;
- unlabeled electrical disconnect box;
- reverse polarity on an electrical outlet;
- an inadequate electrical conducting wire.

Hoist Inspection, LOTO, and Guarding Violations

In this next example, the company was cited for not providing proper crane inspections, LOTO policies and guarding violations. The agency made the following findings and proposed penalties of $103,150 based on the following information.

The alleged serious violations for which the employer was cited included:

- failure to maintain overhead hoists;
- failure to maintain below-the-hook lifting devices in safe condition;
- failure to provide procedures and conduct periodic reviews in a program to lock out or tag-out machineryto prevent its accidental start-up during servicing or maintenance;
- failure to keep aisles and passageways clear and in good repair;

- failure to maintain emergency stops on conveyors;
- failure to guard shafts, belts, drive chains, rotating parts, nip pointsand other moving parts on machinery;
- failure to reduce compressed air used for cleaning to below 30 psi;
- failure to store separately oxygen and acetylene cylinders;
- failure to provide appropriate personal protective equipment and training.

The serious violations carry a total proposed penalty of $59,150.

The firm was also cited for alleged other-than-serious violations including failure to label electrical circuit breakers, failure to maintain exit passageways, and failure to provide training and keep material safety data sheets on the hazardous substances in the workplace.

INDEX

W

X